The Penguin Book of Penguins

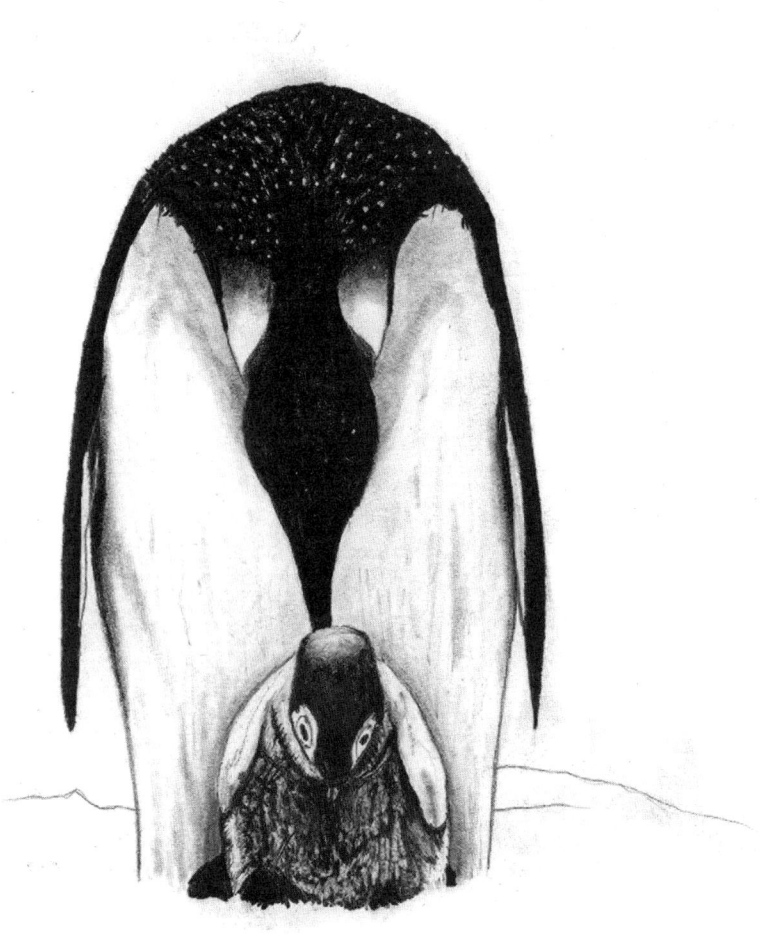
Emperor penguin with chick

THE PENGUIN BOOK OF PENGUINS

An Expert's Guide to the World's Most Beloved Bird

Peter Fretwell and Lisa Fretwell

PENGUIN
VIKING

VIKING

UK | USA | Canada | Ireland | Australia
India | New Zealand | South Africa

Viking is part of the Penguin Random House group of companies
whose addresses can be found at global.penguinrandomhouse.com

Penguin Random House UK,
One Embassy Gardens, 8 Viaduct Gardens, London SW11 7BW

penguin.co.uk

First published 2025
002

Copyright © Peter Fretwell and Lisa Fretwell, 2025

The moral right of the authors has been asserted

Penguin Random House values and supports copyright.
Copyright fuels creativity, encourages diverse voices, promotes freedom
of expression and supports a vibrant culture. Thank you for purchasing
an authorized edition of this book and for respecting intellectual property
laws by not reproducing, scanning or distributing any part of it by any
means without permission. You are supporting authors and enabling
Penguin Random House to continue to publish books for everyone.
No part of this book may be used or reproduced in any manner for the
purpose of training artificial intelligence technologies or systems. In accordance
with Article 4(3) of the DSM Directive 2019/790, Penguin Random House
expressly reserves this work from the text and data mining exception

Set in 12/15pt Garamond MT Pro
Typeset by Six Red Marbles UK, Thetford, Norfolk
Printed and bound in Great Britain by Clays Ltd, Elcograf S.p.A.

The authorized representative in the EEA is Penguin Random House Ireland,
Morrison Chambers, 32 Nassau Street, Dublin D02 YH68

A CIP catalogue record for this book is available from the British Library

ISBN: 978-0-241-73206-9

Penguin Random House is committed to a sustainable future
for our business, our readers and our planet. This book is made from
Forest Stewardship Council® certified paper.

To Holly and Amy

Contents

Preface	ix
Chapter 1: Introduction	1
Chapter 2: Geography and Evolution	10
Chapter 3: Flippers, Feathers and Feet	28
Chapter 4: Behaviour	55
Chapter 5: Penguins and People	96
Chapter 6: Species	130
Chapter 7: Into an Uncertain Future	215
Penguin Jokes	238
Epilogue	240
Penguin Books	246
British Antarctic Survey	250
Further Reading	256
Glossary	259
Acknowledgements	264

Preface

It's been ninety years since the famous, jaunty, black-and-white penguin icon hit the bookshelves. Synonymous with affordable yet quality paperbacks that at the time cost no more than a packet of cigarettes, the perky little penguin emblem became world-famous and launched a publishing empire that has dominated the landscape of book publishing for the best part of a century. Over those nine decades Penguin Books has expanded from affordable paperback fiction to cover a dazzling variety of genres and forms, with thousands of authors and offices around the globe. So, it may come as a surprise that Penguin, a brand so synonymous with that black-and-white icon, has never published a *Penguin Book of Penguins*. So, when someone came to me and asked if I would like to write *The Penguin Book of Penguins*, it was really something that I could not say no to. I am fortunate to have worked on and around penguins for over twenty years at British Antarctic Survey and in that time I have been able to make some groundbreaking discoveries that have helped change our understanding of these little birds. Most of my work has been on emperor penguins, discovering and documenting their populations and charting their battle against climate change, but I have also published on a number of other species of penguin and, having spent many field seasons in Antarctica, I have visited dozens of penguin colonies. If you see so many penguins and spend a lot of your time researching them, it is difficult not to develop an affinity for all things penguin and to absorb endless amounts of penguin trivia, facts and folklore. Working in such a well-known polar institution, knowledge and learning on all things about Antarctica seems to seep into your brain almost by osmosis. So, I knew that writing this book was an opportunity not to be missed.

And what a subject. Penguins are unique. They have charm, charisma and a star quality beyond all other birds. How many other birds have had a dozen Hollywood movies made about them? Or have so many icons, brands and memorable characters (remember Mumble, Pingu or Feathers McGraw?). In real life they can be funny, curious, cute and beautifully elegant. They have unique abilities and have amazing, sometimes bizarre, adaptations that have enabled them to conquer some of the most extreme and harsh environments on the planet. This book reflects that uniqueness and celebrates the idiosyncrasies and special qualities of this incredible bird, recording all of the well-known strange abilities and lesser-known, but incredible, attributes that make penguins so special. The book also relays their relationship with humans, a relationship that has been as rocky as the wind-swept islands on which they breed. Since we discovered these curious, flightless birds we have loved them. We have also eaten them, stolen their eggs, plucked them for their feathers, squashed them for their oil, evicted them for their guano, bulldozed their habitat, and overfished the seas until no food remains for them.

In the past two centuries their numbers have plummeted until today, when penguins are the second most endangered of all bird families and several species are on the verge of extinction. It seems that everywhere humans have been they have had a detrimental impact on penguins. It was only in Antarctica, where there are no humans, that it seemed like penguins were safe, but now with the spectre of global climate change even the remote, pristine wilderness of the far south is changing and the habitat for many species of penguins is becoming increasingly untenable.

Sometimes, when working on penguins and relaying the story of their seemingly hopeless plight, it can be easy to feel depressed and saddened by what is happening. But there is hope. This book also highlights some of the fantastic work done by volunteers and conservation bodies that help to save penguins, bringing them back from the brink. Unlike many other rare island-living animals, over the last two centuries, no species of penguin has gone extinct, although some are teetering on the precipice. But their future is still not written, and we have to believe that there is time to save all of them.

This book is a celebration of penguins, to highlight our love of and wonder at this most remarkable of birds and hopefully to highlight their predicament and struggles in our modern world. And to remind us, *there is still time. Just.*

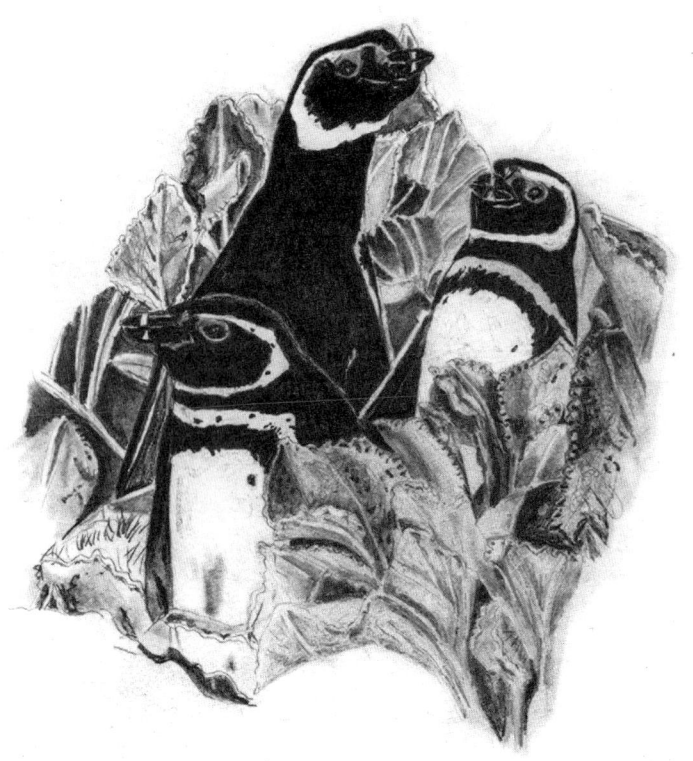

Magellanic penguins on the Falkland Islands

Chapter 1:
INTRODUCTION

There is something about penguins. They are certainly one of the most charismatic of all animals, adored and appreciated by almost everyone. But it is hard to put your finger on exactly what it is that gives them this elevated status. Maybe it is their comical waddle, or the fact that their upright gait makes it easy to compare them to humans. Their clumsy manner on land, social habits and friendly nature endear and fascinate us. We love to anthropomorphize them, often comparing them to old men or waiters wearing a smart suit. Black and white, sharply dressed and unique, their appearance makes them instantly recognizable. Different species can be exotic, rare, elegant or funny and the fact that many of them have extreme lifestyles makes them even more interesting. They have a host of unique characteristics and adaptations that enable them to cope with some of the harshest environments on earth. So, they are not only cute, but scientifically fascinating. Whether it is their ability to withstand the extreme cold of Antarctica, or to dive down to ridiculous depths in search of food, or their sometimes bizarre breeding habits, these southern birds have fascinated us for centuries. The charm of this small, black-and-white bird has transferred into popular culture, through films, TV and documentaries. Penguins have become ubiquitous in advertising, used to symbolize cold temperatures or for their comic nature as well as the endurance and hardiness of this extreme survivor.

For most people living in the northern hemisphere, they will never see penguins in the wild and will only ever encounter them in zoos or on the screen. However, like many other charismatic species, this does not distract from their popularity. Even in a zoo they ooze charisma and charm.

If you are ever lucky enough to get to a penguin colony, especially one of the really big ones in the Southern Ocean or Antarctica, you will find an experience very different from what you may get in the zoo back home. Penguin colonies are massive, often numbering in the tens, or hundreds, of thousands strong. I have visited many of these huge metropolises and each visit is a truly memorable experience. They are a cacophony of noise and action: a huge mass of life that can stretch miles, and the smell, well, that is something that the David Attenborough documentaries don't tell you about. If you stay there long enough it feels like the acrid odour will dissolve the skin on the back of your throat. But it's worth it. The sight of these charismatic animals in such numbers, for the most part totally unfazed by humans, is a sight that will stay with you for a lifetime.

I have studied penguins along with other polar wildlife for over twenty years and I know from experience that penguins are different. When we publish a science paper or release a media story about any of the animals that frequent the frozen continent, we always get press interest. Whether the story concerns whales, seals or albatrosses, we can guarantee that many people will engage and be intrigued by them. But when we release a science story on penguins, the interest is at a different level of magnitude. The recent science stories that I have been involved with on emperor penguins and climate change reach areas and touch people in a way that virtually no other species can.

Press releases often go global on traditional and social media, breaking down the barriers to communication that more highbrow science about physical climate change or sea-level rises struggle to overcome. And that is one of the reasons that I study penguins and have spent much of my career researching them. They are a fantastic mechanism to get across a message about the importance of Antarctica and the Southern Ocean: a window to the ecosystem and physical processes that control the wild and extreme environment that surrounds the southernmost continent. A place that is hazardous and remote, but also connected to all of us in the Earth System. Many of the distant places penguins inhabit face changes and threats to their existence through human interaction, habitat loss, invasive species and climate change. They may be far away and isolated, but the world they inhabit and the changes they face are interconnected and will be felt by us all, to a greater or lesser degree, in future decades.

Did I also mention that I really like penguins? As a scientist you are not really supposed to have a favourite species, even a favourite type of animal. So I must be a bad scientist, because I really like penguins!

Of course, I am not the only person affected by this

flamboyant creature; I'm just lucky that it has become my job and privilege to go to places and see the birds close up. Another admirer of penguins, the great Victorian critic and writer John Ruskin, liked them for a different reason – they cheered him up! In a famous letter to a friend in 1860 he wrote, *'When I begin to think at all, I get into such states of disgust and fury at the way the mob is going on that I choke; and have to go to the British Museum and look at Penguins till I get cool. I find Penguins at present the only comfort in life. One feels everything in the world so sympathetically ridiculous, one can't be angry when one looks at a Penguin.'* Goodness knows how Ruskin would have felt if he'd seen a real penguin colony, rather than a few tatty stuffed ones in a museum!

In our imagination penguins inhabit the icy wastes of Antarctica, but in truth only four of the eighteen species are truly Antarctic. Most of the others live in the cool temperate climes of the Southern Ocean, and several species exist in much warmer environments, as far north as the equator (a few actually breed just over the line in the northern hemisphere). As a family, penguins have many unusual adaptations; their plumage and bodies are uniquely adapted to a life in the water. They lost the ability to fly around 65 million years ago, believed to have evolved from a single ancestor adapting through time to different conditions and challenges, witnessed through many fossil species, to the birds we see today. They have developed wings that can hardly bend, with plate-like feathers and extremely short legs, which give them that waddling gait.

Each species has its quirks, its unique features and its own challenges. Take for example the tiny little blue penguin

INTRODUCTION

in New Zealand and Australia. It often lives within spitting distance of humans and is threatened with extinction from invasive species and climate change, but it has been helped massively by conservationists who make penguin homes and nest boxes to try to protect the birds from predators. Contrast that to the remote wilderness home of the emperors, far away in the icy wastes of Antarctica where less than half of all the colonies have ever even been visited by humans, but are still at risk in a warming world.

To live in these habitats, penguins have developed a range of unique behaviours and adaptations. One of the things that has amazed me while writing this book has been the range of adaptations their bodies have developed, some ingenious and others just downright weird. The more unique the habitat the more unusual the adaptations, so it is no surprise that the emperor penguins seem to hold many of the records. They have the warmest feathers, are the deepest divers, the tallest, heaviest and most southerly breeding penguin, and the only warm-blooded animal to breed in the Antarctic winter. To do this they have developed incredible bodies, behaviours and strategies. But there are many other strange traits and peculiarities across almost all penguin species.

The emperor's cousins, the king penguins, have a strange, multi-year breeding cycle, with the chicks spending a year and a half maturing before they fledge their brown baby feathers and get their smart black-and-white suits.

It's not just the imperious kings and emperors that exhibit remarkable behaviours. The smaller penguins have interesting social habits too: complex courtship displays, dependable parenting and lifelong partnerships can turn to a darker side of fighting, stone stealing and polygamy.

Gentoo courtship display

The raging storms of their ocean habitat have made these smaller penguins tough. Often bashed and tumbled by the tempests surrounding their island homes, they run the risks of waves, rocks and cliffs to reach their nests. They also have many predators. In the sea, leopard seals and orcas feast on them, while in warmer waters sharks and a range of other predators abound. On land, avian predators such as skuas and giant petrels are the enemies in the south, while in the more temperate regions, many other types of predators such as gulls and hawks target these species. And yet, with all these challenges and hardships, penguins abound. There are tens of millions of them around the Southern Ocean. Single colonies can be over 1 million strong, and overall it is thought that the total penguin population numbers around 42 million.

The actual number of species is hotly debated amongst scientists, with some types thought to be mere subspecies, while other scientists argue that some forms currently regarded as subspecies are in fact separate species. Even in this age of DNA and genetic analysis, the arguments persist.

Currently, the family of *Spheniscidae* is split into six genera: the two great Penguin species, the kings and emperors, form the *Aptenodytes* genus; the other three Antarctic species, the gentoo, Adélie and chinstrap, form the *Pygoscelis* or brush-tailed genus. The four species of banded penguins or *Spheniscus* genus live further north than other penguins, venturing into the heat of the Galapagos Islands, Chile and South Africa. The largest genus is that of the crested penguins, of which there are seven species, mostly living in the turbulent Southern Ocean, often on small, remote, windswept rocky islands. This leaves the two odd-ones-out, the little and the yellow-eyed species, each the sole representative of its own genus. Every species is different, with many instantly identifiable, although some are annoyingly difficult to tell apart unless you know where they have come from.

People and penguins have shared a rocky relationship. Like the dodo, their flightless nature made them easy prey for early human colonizers. Polynesian settlers are thought to have driven the Chatham Island penguin to extinction in the 1500s and it is lucky that, so far, no other species has gone the same way. It's not like humans haven't tried. Hunting, in the sealing and whaling era, led to the collapse of many populations. King penguins were particularly targeted, with gruesome inventions like the Penguin Digester – a boiler which rendered the fat from tens of thousands of birds on Macquarie Island in the Southern Ocean. The introduction of invasive species like cats, rats and possums drove many ground-nesting bird species like penguins off the larger islands, only finding sanctuary on the wave-washed islands too small, remote or wild for humans to colonize. But this has not saved them from overfishing, pollution and the looming threat of climate change. Each species has its own challenges, but most are in danger, with thirteen of the eighteen species listed as threatened or near threatened or worse on the International Union for Conservation of Nature (IUCN) Red List. Overall, two-thirds of penguin species are considered threatened, vulnerable or endangered, with one, the African penguin, classed as critically endangered – the last stop before extinction.

This book covers these topics, from what makes a penguin a penguin, to their geography and distribution, to their biology and behaviour, as well as a history of human interaction and the threats to their future. It also covers the lighter side of penguins, a celebration of our love affair with them, highlighting some of the cultural references, characters and jokes about penguins, along with a number of the more interesting anecdotes and stories I have found while working alongside our little black-and-white friends for over twenty years.

A plethora of penguins.
Various versions of the logo used by Penguin Books.

Chapter 2:
GEOGRAPHY AND EVOLUTION

What is a penguin?
Let's start with the basics and define what a penguin is. Penguins are flightless birds that spend most of their lives foraging for food in the ocean but (mostly) come on shore to breed. So, they are aquatic and classed as seabirds. In the past sailors and explorers have wrongly classified them as fish (because they live in water) or mammals (as their feathers are so fine they resemble fur), but they are most definitely birds. In taxonomy terms they form part of their own order and family and are quite separate from all other birds.

They (mostly) breed in large colonies and all (well almost all) live in the southern hemisphere. From a distance they are black and white, although many species have splashes of more extravagant colour. Unlike other birds their wings are stiff and blade-like and they (usually) have an upright gait to help with underwater aerodynamics. As mentioned in the introduction there are currently eighteen species of penguin (hotly debated), many of which are under threat of extinction.

As you can see from the above definition there are a lot of generalities in our understanding of penguins. It is the details and exceptions that make the most intriguing stories and we will expand on all of them over the coming pages.

Penguin populations

There are lots of penguins in the world. From the eighteen officially recognized species, almost all of the 40 million-plus adults live in the southern hemisphere, with just a few hundred individuals venturing north of the equator. The most populous penguin is thought to be the macaroni, which lives in vast numbers in a huge array of breeding colonies on a number of sub-Antarctic islands. Indeed, the sub-Antarctic islands, those rocky fly-specks of land that ring the Southern Ocean, are hugely important for many seabirds and over half of all penguins breed on them. The lion's share of the other penguins exist around the Antarctic coastline, where around 15 million adults turn up annually to raise their chicks. Some southern hemisphere countries like Australia and Argentina still have millions of penguins breeding on their shores, but several others such as South Africa, Namibia, Peru and Chile that once had many millions, and in some cases tens of millions, of happy birds have, over the past few centuries, been devastated by industrial levels of exploitation and those populations have now dwindled to just a few thousand.

Although Antarctica may not numerically have the

most penguins, it has become synonymous with the birds and it's true to say that on the barren wilderness of the white continent penguins have become the dominant and most visible life form (some pedants might point out that ice-seals, being larger, constitute a greater biomass than penguins, but these seals breed on the ice around the continent, not on the continent itself).

You may wonder why and how this family of birds has succeeded in this harsh place. The answer is fairly simple: there are no land predators in Antarctica and there never have been, at least not since Antarctica became ice-covered 30 or more million years ago. All superheroes have a nemesis and for our little hero of the bird world that nemesis is the land predator. Cats, dogs, bears, stoats, weasels, rats, possums, armadillos, lizards, snakes and a whole host of other beasts love to eat penguins and penguin eggs. Being flightless and not that agile they are easy prey to sharp claws and strong jaws. So in more temperate latitudes most penguins are limited to breeding on tiny islands where things on four legs can't get to them. But in Antarctica none of these land-based predators exist and this has enabled a superb aquatic hunter like the penguin to thrive. Over time they have adapted to tolerate the extreme conditions and exploit the rich and productive waters that surround the ice-bound continent.

When I told my footballing friends down the pub that I was writing a book on penguins, they started quizzing me about penguin facts. Most of their questions I answered fairly easily, until someone asked me,

'Which country has the most penguins?' The answer was one I was not immediately sure of, and the solution was not easy to find. You may think that the simple answer would be Antarctica, but the Antarctic is not a nation state, rather a continent ruled over by an international committee (based on the Antarctic Treaty). While the other southern hemisphere nations do have penguins, most live on those sub-Antarctic islands, which are often dependencies of European nations, especially the UK and France. So, I set to work and added up all the numbers listed in the Birdlife International records and the result was quite surprising. Of all the countries in the world, France has jurisdiction over the most penguins, with around 10 million living in its territories, mainly the islands of Kerguelen and the Crozet Archipelago. Second is the UK, which hosts around 7.5 million, spread between the Falkland Islands, South Georgia, the South Sandwich Islands and Tristan da Cunha. Australia has about 5 million and Argentina about 2. Australia does however have the most species breeding in its territories (seven different types), with the UK and New Zealand both on six species.

But that is not quite the end of the story. While Antarctica is governed by the Antarctic Treaty, which has successfully endured since 1961, the old powers and several southern hemisphere countries still have outstanding claims on its territory. The UK, along with Argentina and Chile, all claim the Antarctic Peninsula, Australia claims East Antarctica and New Zealand has planted its stake in the Ross Sea sector. The territorial claims in Antarctica are still there, on hold under the Treaty. So if we add these in then the UK has jurisdiction over the most penguins as it claims the sector that includes the Antarctic Peninsula, which includes some of the largest mega-colonies on the continent.

Before we go too much further into this book it is worth briefly listing the types and species of penguins. In Chapter 6 each species is described individually, but here we will just go over the basics. Scientifically, the eighteen officially recognized species (I will explain in a moment why I keep saying 'officially recognized') are part of the family *Spheniscidae*, of the order *Sphenisciformes*. These Latin terms come from 'wedge-shaped' and are thought to derive either from the penguins' wedge-shaped bodies or, more commonly, from the shape of their wings. This order can be divided into six types or genera, to be more precise.

The largest genus are the crested penguins, with seven species. Their Latin name is *Eudyptes*, meaning 'good diver'. These penguins are all fairly small in size and are identified easily as they are, as their name suggests, the only type to have a crest on their heads. The crests are the best way to distinguish each species because although all have yellow crests they are different shades, shapes and lengths, depending on the species. Other than their crests, they look fairly similar, and several are quite hard to tell apart. They tend to live on the sub-Antarctic islands and only one species bucks this trend, living on the mainland of New Zealand. The species are:

GEOGRAPHY AND EVOLUTION

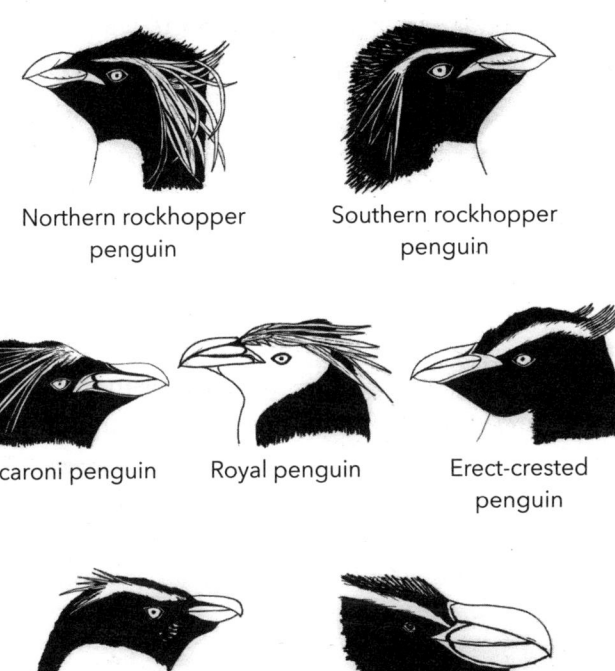

Northern rockhopper penguin

Southern rockhopper penguin

Macaroni penguin

Royal penguin

Erect-crested penguin

Fiordland penguin

Snares penguin

The second-largest genus, with four species, are the banded penguins. The Latin name *Spheniscus* is almost the same as the name for the order of penguins (*Sphenisciformes*) and also means 'wedge-shaped'. They are called banded penguins due to the black band of feathers that surrounds their chests. Apart from that, and a small pink patch of flesh around their beaks, they are purely black and white.

These banded penguins are different in several ways from their crested cousins. Firstly, they live mostly on the mainland of Africa and South America. Here the temperatures are much hotter than the sub-Antarctic, so they dig burrows to get out of the sun and protect their chicks from the heat. Also, unlike the crested penguins, which mostly feed on crustaceans such as krill, the banded penguins prefer fish. They are notoriously hard to tell apart; usually the banding on their chests and the patterns on their heads are the giveaway. The species are:

Humboldt penguin   Galapagos penguin

African penguin   Magellanic penguin

Our third genus, with three species, are the brush-tailed penguins or *Pygoscelis* penguins. This is a rather catch-all group with three easily distinguished species. What links them is that they are all purely black and white (except for their feet and beaks!) and they mainly live in Antarctica. All the brush tails tend to breed on rocky nests in huge, tightly packed colonies often intermingled with each other. They are mainly krill eaters. *Pygoscelis* means 'rump-legged' in Greek, although why they should be called rump legged is a bit of a mystery. I have spent a lot of time with each of these species and I still have no idea why they have this title. The species are:

Adélie, chinstrap and gentoo penguins

Next we have the great penguins, the *Aptenodytes* genus. These are the two largest species, the king and the emperor. Penguin royalty (never mind the diminutive royal penguin!), these two are quite different from the smaller species. For a start they are bigger – much bigger: an emperor can stand almost three times the height of a little penguin and weigh forty times as much. The king, although smaller, still stands at an impressive 1 m tall, twice the height of many other species. They have a different body shape, with long flexible necks and a slender but lengthy bill. They are still predominantly black and white, but rather than a yellow crest or black chest bands, they have golden-yellow cheek patches that make them very easy to identify. Like the other species they eat fish, squid and krill, often with a preference for fish, which they dive to incredible depths to catch. To go so deep and hunt in the icy blackness of the oceans for long periods they have both developed a raft of specialized adaptations. The other thing that sets them apart is their strange, some would say bizarre, breeding habits. Kings and emperors both have very interesting strategies to raise their chicks. Because they are both big birds their young are too large to reach maturity over the short polar summer and they have each developed very different approaches to getting around this problem.

The last two species of penguin both have a genus of their own. The yellow-eyed penguin is the size of a gentoo (the third-largest penguin) but has a mostly yellow head, along with that yellow eye, and it is probably a close relative of the crested penguin. The Latin name of its genus is *Megadyptes,* meaning 'large diver'. The last genus is that of the little penguin, although it has many other names, depending on where you see it. Luckily the Latin name is always the same, *Eudyptula*, referring to it as a 'good diver'. As its name suggests, it is the smallest penguin, and it is quite different in shape, behaviour and character from the other species.

So, there you have it. Eighteen species of our well-loved black-and-white feathered friends. There once were more and recently there were less . . . but we will come on to that in a moment.

The evolution of penguins

Penguins have a long evolutionary history. They separated from other seabirds as long as 65 million years ago, just as the age of the dinosaurs was coming to an end. The unique features of this early flightless seabird, which used its wings for propulsion rather than its feet like most other seabirds, can first be seen in the fossil record from around this time.

GEOGRAPHY AND EVOLUTION

Although we think of penguins as polar animals, adapted to the cold, Antarctica is not where they first evolved. That honour goes to New Zealand, or at least the landmass that geologists call Zealandia, which over the eons developed into the islands we now call New Zealand.

Millions of years ago this landmass was much larger and much further east in the Proto-Pacific Ocean, separated from the coasts of Australia by a much greater distance. At that time, Australia and most of Antarctica were part of a larger continent called Gondwana. The much smaller mass of Zealandia had spilt apart from the other continents over 100 million years earlier, and one consequence of this was that mammals that had started to evolve after Zealandia had broken away never had the chance to colonize the remote continental plate. Of course, the dinosaurs and other great lizards were there, and birds had managed to fly over the ocean to settle on the isolated landmass. But 65 million years ago, at the end of Cretaceous period, when that infamous meteorite hit the earth and the dinosaurs went extinct, suddenly there were no land-based predators in Zealandia to eat them, and no herbivores to compete for food. The remaining animals found that there were many evolutionary niches that had been left vacant and needed to be filled, and over the next few million years they quickly adapted to fill them. On the other continents, mammals evolved to fill these gaps and became the dominant land animals, but in Zealandia there were no mammals, so over time birds evolved to occupy these roles. Without any terrestrial predators, there was no need to fly, so many of them became flightless, like the giant extinct Moa bird and many species that still exist on the islands. And it wasn't just on the land that birds evolved to fill the vacant ecological niches; in the sea, without the huge aquatic lizard predators that also went extinct with the dinosaurs, flightless seabirds evolved in Zealandia to take up the mantle of marine predators, and it is these that are the ancient ancestors of today's penguins. With no further need to fly to escape from hunters, their wings were redundant on land, but in the sea they developed into flippers, a much better propulsion system than feet, which most other seabirds use, giving penguins a real evolutionary advantage.

Some of these early penguins were truly huge. The largest yet found, *Kumimanu fordycei*, lived on the ancient beaches of Zealandia 57 million years ago and weighed in at 150 kg, three times heavier than the largest penguin living today. There have been several early fossil penguins found in the rocks of modern-day New Zealand, but such a successful bird didn't just stay on its homeland and soon started to colonize the landmasses nearest to it. At that time, Zealandia was closer to the Antarctic Peninsula and the west coast of South America and it is these locations where we start to see the next generations of fossils turning up in the geological record. In those times South America was joined to the Antarctic Peninsula, forming a blockage so that penguins could not access the Atlantic or Indian Oceans. However, when the Antarctic Peninsula broke away from South America about 48 million years ago, the ancestors of our little friends could expand into all of the Southern Ocean. It was boom time for penguins, and a multitude of species, many more than there are today, evolved. One of these was *Palaeeudyptes klekowskii*, also known as the colossal penguin, a species found in the fossil record of the Antarctic Peninsula that lived 40 million years ago and stood over 2 m tall. It had a fearsomely long beak, full of razor-sharp teeth for catching fish, and must have been a truly impressive and probably quite frightening sight!

The next big change was in Antarctica. With the South America/Peninsula split, the Drake Passage started to open up and, as this widened, the Circumpolar Current started to develop. Over time this became the strongest current in the world and formed a barrier to weather systems, meaning that warmer air from the tropics was blocked from reaching the Antarctic continent. The Antarctic, which had previously been much more temperate than it is today, started to get cold and ice sheets started to form. The trees died, the land became frigid and the ice advanced. Before this, not only was the southernmost continent green, but it was also not a single landmass, rather a number of large islands, with lots of available nesting coastlines for penguins. But as the ice relentlessly advanced over the next few million years, it covered the land and thickened until it was 2 km deep, linking those islands into the single landmass we know today as Antarctica. Most of the really large penguins died off, but there were still lots of different smaller species spread across the Southern Ocean. It was during this period, around 8 million years ago, that the great penguins split off from the smaller species and started to develop into the two large penguins we see today. The other penguin genera started to develop too, and the groups of banded, brush tail and crested can all be seen to originate over the last few million years.

Our next landmark in penguin evolution came around

2.5 million years ago. These were the ice ages. Geologists are not entirely sure why the earth started to experience ice ages at this time. Maybe it was the locking away of too much carbon deposited over time in rocks, or maybe some unknown geological event tipped the earth over the edge, but about this time the world's climate started to slowly oscillate between warm and extremely cold periods. It is a time that geologists call the Quaternary period. Cold glacial periods lasting around 100 thousand years were interceded by warm inter-glacial periods, when temperatures were often warmer than they are today. These inter-glacial periods tended to be shorter, around 10 to 30 thousand years. The whole cycle was ruled by the earth's orbit and small deviations in the tilt and wobble of how the planet went around the sun.

Whatever caused the ice ages and ruled over its periodicity, it had a profound effect on the evolution of many species, including penguins. Many species died off, unable to cope with the extreme cold, but others adapted and evolved to take their place. These natural changes to the climate were fast in geological terms, but still took thousands of years to switch between cold and warm, enough time for penguins and many other animals to evolve to cope with the new challenges. During the glacial periods, the Antarctic ice sheet and the sea ice would extend into the Southern Ocean and many of the sub-Antarctic islands would have their own ice caps. Throughout the warm periods, sea ice would retreat back until only a few areas in the very southern extremities of the Ross and Weddell Seas would have stable sea ice. It was under these dynamic conditions that today's penguins evolved. Some species adapted to the extreme cold of Antarctica, others to the wild weather of the Southern Ocean, and still more to the hotter coasts of the southern continents.

So, penguins evolved to fill many ecological niches in the rich waters of the southern hemisphere. One question that I often get asked is, 'If penguins evolved to cope with the inter-glacials, can they evolve again to beat the challenges of climate change?' But something that genetics and palaeontology has discovered is that penguins have a significantly slower rate of genetic mutation than other birds, making them the slowest to evolve of almost any warm-blooded animal. And, even though they have adapted to climatic variations in the past, those changes took place over thousands of years. Today's human-induced climate change is taking place over decades, not millennia, so there just isn't time for animals to adapt to it, especially not penguins.

Splitters and clumpers

The question of exactly how many penguin species there are is a thorny one. You would think that it should be easy; each different type should be a species, but for birds that live on isolated islands, often there are many slight variations in size, body shape or colouration. In some species, nesting habits and other behaviours differ too, so where do you draw the line on what is and isn't a species, and what is a subspecies?

In this book, I have referred to the 'official' number of penguin species as eighteen. 'Official' probably isn't quite the right word, but there are eighteen species recognized by the main international avian conservation bodies, the

International Union for Conservation of Nature (IUCN) and Birdlife International. Both list eighteen species, but both authorities note that there is some difference of opinion and this list might change.

Taxonomy, the science of classifying and describing animals, first began when Carl Linnaeus, a Swedish biologist, famously invented his system of classification in the 18th century. Throughout the following century there was a huge craze for collecting and classifying animals and plants, but there has never been a real agreement on what makes a species, or how different something should be before it becomes one. If you look across different animal groups there are huge disparities. Birds in the northern hemisphere are often classed at species level with only minute differences in colour or behaviour, whereas many other families of animals with slightly different characteristics, especially in the ocean or remote locations, are bundled together as one species. If you take the example of the killer whale, there is only one recognized species, even though the different subspecies (usually referred to as ecotypes) differ in size, shape, colouration, behaviour and diet. A 'Type A' Antarctic killer whale can be twice the size and different in colour, patterning and diet from the smaller Ross Sea, 'Type C' killer whale.

It was believed that, when the revolution in genomics came along and we could 'barcode' each individual species, it would be the saviour of taxonomy. But if anything, being able to look in detail at each genome only muddied the waters even more. As we started to code the different gene sequences of animals it became clear that differences in the genetic code and appearance did not always line up.

GEOGRAPHY AND EVOLUTION

Take the household dog for instance. As we have only been breeding domestic hounds for a few thousand years, the genetic code across different breeds of dog is very similar, but their appearance can vary wildly. It is the same for many animals, including birds and penguins. Genomics tends to give clues about the mixing and evolution of animals, but it is not always that good at classifying what we think a separate species is.

When I started working on penguins twenty years ago, there were seventeen recognized species. Today there is officially one more type: the rockhopper penguin was originally considered a single species, but it was split into Northern (*Eudyptes moseleyi*) and Southern (*Eudyptes chrysocome*) species in 2006. These two species had obvious differences in the size and shape of the crests on their head. Many other scientists also recognize an Eastern (*Eudyptes chrysocome filholi*) type of this widely distributed bird, but as yet that is not 'officially recognized'. This (currently) subspecies breeds far to the east of the southern rockhopper and has a fleshy pink lining around its bill. It is quite possible that in future this subspecies may also be officially recognized. So how different does a subspecies have to be before it is classed as a separate species and why does it matter?

Well, one of the reasons that it matters is because of conservation. If a species is under threat of extinction, it is a big deal. Conservation efforts can be more easily applied, funds can be raised, political and public pressure can be asserted to try to alleviate the threats to that species. But if it is only a subspecies or, worse, just a regional branch of a larger population, it is much harder to justify that effort and to get people, politicians and policymakers excited enough to make the changes needed to make a difference to their survival. So, what is, or isn't, a full, bona fide species is often a point of contention and hotly debated amongst scientists and conservation bodies.

A good example is the white-flippered penguin. This is currently considered a subspecies of the little penguin, but it has several morphological differences. It is slightly larger and is more grey than blue in colour, and its flippers have a white band around their edges. It is, overall, quite different from its little blue relatives, and most Australian and New Zealand authorities consider it a different species, but it is not recognized as one by the IUCN or Birdlife International. If it was, then it would immediately become one of the world's most endangered penguins, as there are only just over 2,000 pairs in existence and their breeding range is extremely limited, which would both be red flags for conservation bodies. I guess that one of the problems is that there are several other regional differences in the population of the little penguin. The Australian and New Zealand populations are slightly different, and there are variances in breeding behaviour in different parts of Australia. So, if you did split the little blue penguin and the white-flippered, would you also have to make the other subspecies separate species, and where would you stop?

Scientists are divided on the subject and generally fall into two camps. Those who would like to see more species are colloquially classified as 'splitters', and those who would like less division and to keep more subspecies are often referred to as 'clumpers'.

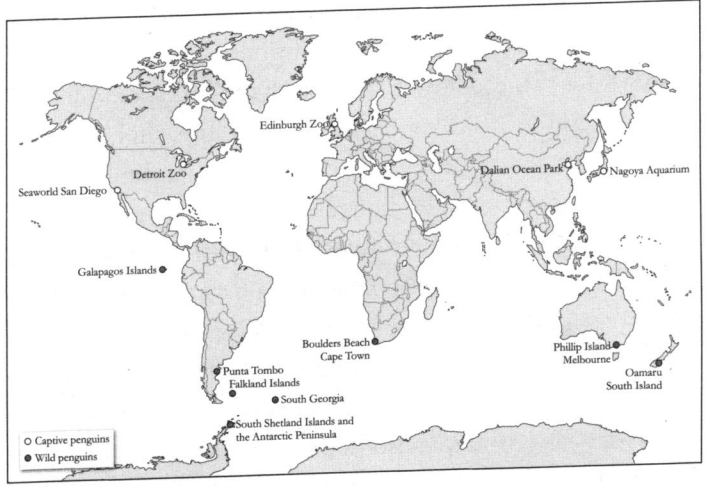

There are many possible places where you can see penguins, but they generally divide into captive or wild penguins. Many hundreds of zoos, wildlife parks and aquariums keep penguins. Some zoos have several species and some have penguin history. I have highlighted on the map several places that are particularly notable, either for the species they keep, or their extensive penguin enclosures. If you would like to visit penguins on their home turf you may have to go further afield. Some species live near civilization in Africa, South America or Oceania and are fairly easy to get to. But to see polar species you will generally have to take a cruise ship to South Georgia or the Antarctic Peninsula.

Chapter 3:
FLIPPERS, FEATHERS AND FEET

Bodies

The penguin's body is unique, and most of that is due to their aquatic lifestyle. Although they appear clumsy on land, these birds are brilliantly adapted to exist in the water. Over time they have evolved to optimize their ability to hunt and forage underwater, often at the cost of their aptitude to live in the terrestrial environment. Not only do they catch their food at sea, but most of their predators are also found under the water, so penguins need to be fast and agile swimmers. Several of the species also have to cope with the extreme cold of the Antarctic and some of the most dangerous conditions tolerated by any warm-blooded species. To do this they have developed a number of unusual biological adaptations; it's worth investigating a few of the more interesting ones.

Unlike other birds they stand erect. It is one of the obvious tell-tale signs that a penguin is a penguin and, thinking about it, there are few, if any, other animals that have such an upright gait. It is probably this erect stance that we relate to and compare to that other upright creature – humans – and is why we love to characterize penguins like little people.

The species range in height from the emperor, standing up to 1.2 m tall, to the aptly named little penguin at a diminutive 33 cm. Emperors are by far the heaviest and can weigh in at a whopping 40 kg, with some individuals occasionally tipping the scales at 46 kg – the same weight as a teenager. This makes it the fourth-heaviest bird in the animal kingdom, after its flightless cousins the ostrich, cassowary and emu.

To reduce drag when swimming, their bodies tend to be torpedo- or spindle-shaped underwater, a much more aerodynamic and powerful shape than that of diving birds like the guillemots, razorbills and their other northern counterparts. Their bulky torsos are made up mainly of chest muscles, which power their flippers underwater. The emperor penguin, which is commonly considered the fattest of all penguins, has the largest chest muscles of any animal compared to its size.

This efficient shape and their powerful muscles mean that they are agile and superb swimmers, more agile than most predators and diving deeper than any other seabird.

Galapagos penguin swimming

Feet and legs

Here in the UK when it gets icy, our national health service often suggests that people take short steps and 'waddle like a penguin' so that they don't fall over. It's true that taking short steps does help us, and penguins, balance in icy conditions. But this is not the reason penguins waddle. They don't waddle by choice. The true reason they have developed such a comic walk is because they have extremely short leg bones that are set far back in their abdomen, which gives them their upright gait. Their lower leg bones are the shortest by relative size of any vertebrate animal; in the water, legs are redundant; penguins use their flippers to swim, and don't flap their feet like a diving duck does, so the shorter their legs are, the better. Add to this the fact that, in Antarctica, you need to keep legs and feet warm, and without the feathers that cover the rest of the body, this can be a challenge. The answer has been to evolve short legs so that the birds can sit on them and their body feathers can keep legs and feet insulated from the cold.

The emperors, which survive in the coldest conditions, have to go to even greater lengths to keep their feet warm. In the Antarctic winter, temperatures get down to −40°C and below, and the penguins have to stand on the frozen ice for long periods of time, which sucks warmth from the soles of their feet. To minimize contact with the snow, adult emperors have a bony heel which they rest on. When it gets really cold, the birds rock back on to their heels and raise the front of their feet up off of the snow, only keeping contact with a very small bit of their feet. Supported by their stiff tail feathers, they form a sort of tripod, a stance that keeps their feet away from the warmth-sapping ice and helps the emperor conserve heat. King penguins also have this adaptation and it has the added benefit of forming a convenient shelf to rest their egg on while brooding.

Tripod stance

All penguins have strong claws, which they use for movement and climbing, a useful adaptation when you have to climb steep cliffs and jump from rock to rock like the acrobatic rockhoppers. Some burrowing penguins also use their claws for digging their burrows, which can go many metres underground. Penguins that live in warmer climes seem to have relatively longer legs, as their need to conserve heat is less severe.

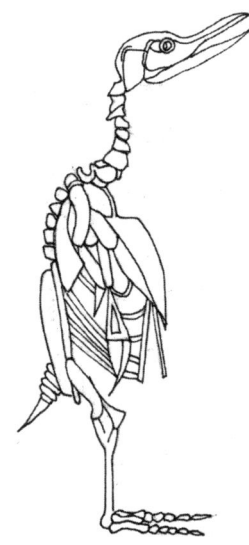

Anatomy of a penguin

Flippers

Penguins are the only bird that cannot really bend their wings in the middle. The appendage is not really a wing at all, and could be better described as a flipper. These flippers are curved like scimitars to minimize drag in the water and are firm and strong, so strong that when adults tussle, they often fight by flipper bashing each other.

They use their huge chest muscles to power these flippers under the water and it is often said that while penguins cannot fly in the air, they fly beneath the waves. When I was first doing fieldwork on emperor penguins, I was warned by an experienced scientist that being whacked by an emperor's flipper could break your arm. Indeed, handling emperor penguins can be a risky business and you can come back

FLIPPERS, FEATHERS AND FEET

Flipper bashing

with a plethora of bruises. Flippers are covered in short, scale-like feathers, all of a similar size on the outward-facing side of the wing, although they often have less dense feathers on the bottom of the flipper that can help with heat loss when overheating. When it gets cold, flippers are held close to the body to keep the animals warm. Many penguins use their wings to stabilize themselves. Their upright stance gives them a high centre of gravity, so like a tightrope walker most penguins will hold their flippers out to help them balance, especially when running or scuttling. The exception, once again, is the emperor penguin. In Antarctica it's too cold to hold out your wings for long periods, so an emperor walks with its wings tucked tight to its body. This is a reliable way to distinguish between an emperor and a king penguin as a king will almost always walk with its flippers held out from its body.

Feathers

The feathers on a penguin are extremely dense and, unlike other birds, almost all of them are short and compact, so much so that in the past people thought they had fur. But they are feathers and, although short, they are not all the same. Penguins have four main types of feather, although scientists have identified over twenty different specialized feather types in emperor penguins. Here we will stick to the main four: the wing feathers, tail feathers, contour feathers and adult downy feathers.

Unlike other birds that often have several different types and lengths of feathers on their wings, the feathers on penguin wings are all short and stiff; they were so unusual that early explorers thought they were scales. Without the need to fly, flight feathers have disappeared and over time these feathers have reduced in size to encase the flipper-like wings with a thin, low-drag covering optimized for fast swimming. The tail feathers are the longest of the four types but are stiff and often used for balance and leaning on. They are longest in the brush-tailed penguins: the chinstrap, Adélie and gentoo, which use them to lean upon and balance on the ice.

The main body of the penguin has two types of feathers: the contour feathers and the adult down feathers. The contour feathers are longer and give the bird its shape, its waterproof protection and its colour. Each feather consists of three parts: the bottom part, near the quill, is covered in insulating down feathers. The middle part is more structured, with filaments like a standard bird feather; these form a protective barrier, keeping the water away from the skin. This part also gives the penguin's body its colour: black on the back and white on the front. However, there is a third part to the contour feather at the tip, which has a fluffy curved end and is often bright blue. That is why not all penguins are black and white. In the little blue penguin, this blue part is larger, and the effect is that the adults look blue. In the king penguin, this blue part of the feather is also comparatively large, so on close inspection the king seems slate or silvery grey rather than black. Although they look black from a distance, when you get close to emperor penguins you can see that their backs have a beautiful bright blue speck on the end of every feather. Some of these contour feathers will also have a downy plume for extra warmth.

Finally, the adult down feathers are short, dense and fluffy and sit close to the skin between the contour feathers to give the body extra insulation. This unique assemblage of very dense feathers gives penguins unrivalled insulation and is the main physiological adaptation that allows the birds to tough it out in the extreme cold of the southern hemisphere. In Antarctica, the only other animals to live above water in the acute temperatures of the polar night are the seals. Although these mammals have fur, they depend mostly on thick layers of insulating blubber to keep out the cold. In comparison, the Antarctic penguins, the Adélie, chinstrap, gentoo and emperor, have relatively small amounts of blubber and rely very much on the insulating power of their unique feathers, which provide up to 90% of their warmth.

And it is not just the density and variety of feathers that are unusual. Unlike other birds, whose feathers tend to be arranged in rows, penguin feathers are evenly spaced over the whole surface of the body, covering all the skin uniformly, and they provide some of the densest feather coverage of any bird. Each contour feather has its own small, individual muscle which can hold the quill stiff and erect when on land. This has the result of maximizing the amount of trapped air in between the outer part of the feather and the skin, with the smaller, dense downy feathers exaggerating the effect. In the water, the feather can be flattened to form a tight waterproof barrier. The tips of the feathers are kept waterproof when the penguin

King penguin preening

preens itself. A gland just above the tail produces a waxy residue with which the bird covers its plumage to ensure that the feathers are watertight and no moisture can get to the downy filaments near the skin. Penguins frequently spend long periods preening, often as soon as they get out of the water, and, if you have time, it can be quite entertaining watching them stretch and perform contortions to ensure that they cover every patch of their body.

Circulation

Even with the warmest feathers in the world, in Antarctica, sometimes it just gets too cold. The solution for most penguins is either to reduce the amount of heat loss, by various specialized and often slightly weird biological adaptations, or simply to make more heat. Many species of penguin that live in cold climates can increase their core body temperature by shivering. Some penguins, such as the king penguin, can thermoregulate without shivering, increasing their core temperature by releasing enzymes to burn more fat. Another adaptation to the cold is a circulatory system with a counter-current heat exchange. This works a bit like a fridge or heating induction unit. In most animals the distribution of the arteries and veins is separate from one another. But in penguins the arteries going out run alongside or intertwine with the veins, so the warm blood carried from the heart is very close to the cold blood on its way back, all the way down the veins to the extremities. In this way, heat is transferred from the hot blood coming from the heart to the cold blood returning from the skin so that it doesn't have to be reheated in the core of the penguin's body.

This special circulatory piping system saves the penguins significant amounts of heat and energy. Emperor and Adélie penguins can also constrict the arteries in their feet so that they do not get as much blood to them and therefore do not lose valuable warmth when standing on the ice.

Breathing is another specialized adaptation. Similar to the circulatory heat-transfer system that helps penguins from losing heat from their blood, penguins have a breathing heat-exchange system. If you are an emperor penguin and have to breathe air that is -50°C, it could have a dramatic cooling effect. When that cold air got to your lungs it would sap a great deal of heat and would cool you down significantly. So, penguins have special adaptations in their noses to deal with the extremely cold air. There are chambers in their nasal passage that slow down the incoming air. In these passages are mucous membranes, which transfer heat from the warm breath being expelled to the cold air being drawn in. By the time the air gets to the lungs it is already almost at body temperature and virtually none of the birds' precious warmth has to be wasted warming it up.

That is OK for polar birds, but some penguins live in temperate or tropical zones and for them the problem is not the cold but the heat. This is especially true of the banded penguins that live in South America and South Africa. Humboldt, African and Galapagos penguins all live in desert-like conditions and have to endure extreme heat from the tropical sun. When you are covered in insulating feathers and your black back absorbs the sunlight it could obviously pose a significant problem. To make it worse, those insulating feathers make it very hard to get rid of any excess heat. To cope with this the banded penguins have adapted circulatory systems that help them cool down. They have bare faces, with pink fleshy parts that have been developed to help the transfer of heat. Birds cannot sweat, but these bare patches help the birds cool down in the wind and they can increase the blood flow to them if the conditions are right. The main way for these birds to cool down tends to be panting and fluttering their throats. Little penguins don't have these bare pink patches on their faces so increase heat loss through their feet and flippers, pumping more blood into these less well insulated areas when they get hot.

Even Adélies and emperors get hot sometimes. When I have visited the Snow Hill Island emperor colony, the most northerly and warmest emperor rookery in the world, it has been common to see adults and chicks spreadeagled on the ice. It looked like they were sunbathing, but in reality the exact opposite was happening. The birds were increasing the surface area of their bodies that was in contact with the cold snow, which is one of the few ways of cooling down. If it gets even warmer, you will often see these birds sticking their beaks into the snow and breathing through the icy layers to decrease their core temperature.

Colouration

Penguins are black and white, right? Well, not always. Add blue, grey, yellow, orange and pink to that palette of colours if you want the full picture.

The main black-and-white colouration acts as a form of camouflage when they are swimming. Underneath they are white, so when a predator like a leopard seal looks upwards, they see a pale body that blends in with the sky. On their back they are black, so that from above they look like the black sea, although as previously mentioned penguin feathers always have a hint of blue. The amount of blue varies from species to species, with the little blue penguin obviously being the bluest. King penguins also have a fair amount of blue in their feathers, which gives them a silver-grey back. Overall, most penguin researchers classify the king penguin as the most colourful and dapper species.

Of course, there is one other type of feather that some penguins possess that gives certain species their unique character: the exuberant head feathers of the crested penguins. These crest feathers vary in size, arrangement and colour (although they are all shades of yellow) and are the main way of identifying the seven species of crested penguins, from the punk hairstyles of the northern rockhopper, which has the longest crest, to the untidy eyebrows of the royal penguin, which looks a little bit like a bad comb-over. As its name suggests, the erect-crested has an upward-curving crest, the opposite to the Fiordland's brows, which are decidedly droopy. Colour varies from golden yellow in the macaroni, to a more lemon yellow in the rockhoppers, to almost orange in the royal penguin. These wild hairstyles are used for courtship displays, and several of the species can raise or fluff out their crests and will shake and show off these colourful quiffs when trying to attract a mate. The yellow chemical used to make these shades is specific to penguins and found nowhere else in the animal kingdom.

The other main colouration on penguins is the yellow facial markings of the king and emperor penguins. Both have cheek patches, known as auricular patches. These tend to be a very deep orangey-yellow in kings, but a more subtle, paler yellow in emperors. The easiest way to tell kings and emperors apart is to see if this patch is a solid colour; this tells you it's a king, or, if it blends from yellow to white, it indicates it's an emperor. Another colourful penguin is the yellow-eyed, which is unique in having a yellow face and head. A bright yellow stripe runs backwards from the yellow eye and around the head. The rest of the head is a less obvious yellow.

Eyes

Penguin eyes come in a range of different colours. Rockhoppers and macaronis have red eyes, while most of the other crested penguins have brown eyes. Yellow-eyed obviously have yellow eyes, and the banded penguins' eyes can vary between black to brown within each species. Little penguins often have slate-blue eyes. The species that live furthest south, the Adélies and emperors, both have black eyes.

All penguin eyes are highly adapted to see both on land and underwater. This is not an easy thing to do as the difference in densities of air and water means that the two mediums refract light in different ways and, normally, if you could see well in one, you would not be able to focus in the other. This fact means that most animals have to choose between being either fully focused in water, but short-sighted in air, or able to focus in air, but long-sighted in water. To compensate for this, penguins have a special adaptation to the lenses of their eyes which means that they can alter the shape of their cornea, so that it can focus both on land and in water perfectly well. This is another adaptation that is unique in the animal kingdom.

Another ability is seeing in the dark. Penguins dive very deep to catch their prey. Other deep-diving animals that do this, like whales, often use ultrasound, echo location or other vocalizations in the dark depths to find their food, but penguins are visual hunters and have to use their eyes. Light is absorbed in water and the deeper you go the darker it is, so by the time you get a couple of hundred metres down, it is pretty much pitch black. But this absorption is not even across the light spectrum. The longer wavelengths of light, such as red and green, are absorbed first, and shorter wavelengths penetrate deeper into the water. So, many species of penguins can see in ultraviolet to maximize the extreme low light levels in the depths of the ocean. The two deepest-diving species, the kings and emperors, have specialized pupils that can change in size enormously to optimize the low light levels. On land, a king penguin lens can shrink to a tiny square pinhole to avoid snow blindness, but can expand 300-fold when diving in the depths to let more light in. Compare this to humans, who can only expand their iris by a measly 50%; no wonder we need sunglasses in the snow.

Having said that, it is still not fully understood how the really deep-diving penguins such as the emperors manage to catch food at extreme depth, especially in the low light levels of the polar night, and some scientists have suggested that they could hunt by touch or by using the slight bioluminescence of their prey to catch their dinner.

There is some evidence that their eyesight on land is not as well developed as humans'. In Antarctica, at British

Antarctic Survey's Halley Research Station on the Brunt Ice Shelf, they often have a particular problem with emperor penguins. The nearest emperor-penguin colony is located near the coastal resupply site 20 km away. Each year, the station staff set up a line of oil drums from the ice edge back to the research station, marking the route for resupply. When the ship docks at the ice edge, tractors and snowcat vehicles tow fuel and vital supplies across the featureless ice shelf to the research facility. What is strange is that almost every year one or two emperor penguins turn up at the base and have to be returned to the colony by skidoo. They follow the drum line that marks the tractor trail. The common theory is that they mistake the distant drums for penguins and waddle over, only realizing their mistake when they get close. Then they see another drum in the distance and waddle over to that. After 20 km, they end up at Halley!

Beaks

The beaks of penguins are one of the most variable features of the birds. They differ in size, shape and colour. The banded penguins have thick, narrow beaks and each of the four species can be identified by the colour and patterning between the beak and the eye. The king penguin has the longest beak in relation to its size, being 12–13 cm long (about 15% of its height). Its beak is narrow and fairly straight with a pronounced downward curve at the end. It has a bright orange plate on the lower beak called a mandibular plate. Emperor penguins also have this plate, although it's more pink in colour and not as large or prominent as the king's. The crested penguins all tend to have pink or orange beaks, while the banded penguins and some of the brush-tailed species have mainly black beaks. Generally, the species that eat fish tend to have longer beaks than the krill-eating penguins like the Adélies, whose diet consists almost entirely of krill, a small, shrimp-like crustacean, and the Adélie has the smallest beak of any penguin.

The beaks of all the species have two unusual modifications. No birds have teeth, but when a penguin opens its mouth, you will see that it has something quite similar. Inside its beak, on the inner top and bottom and along the tongue, are rows and rows of sharp, backward-pointing spikes. These can be quite pronounced in the larger penguin species. They act to ensure that prey caught in the mouth of the bird does not get away and it seems unlikely that any fish or sea creature caught in such a fearsome appendage would escape, no matter how much it wriggles.

The second adaptation is salt glands in the nose of the beak. These are most obvious in the banded and crested penguins, which, like albatross and petrels, have special tubes or ridges running up the side of their beaks. These glands enable the birds to distil the salt from seawater, meaning that they can drink the salty brine to rehydrate. The Antarctic species have less obvious salt glands and lack the noticeable ridges seen on the beaks of the banded and crested penguins. Antarctic penguins such as emperors and Adélies often eat snow to rehydrate rather than drinking seawater.

Diet and prey

All penguins forage in the oceans and, as mentioned previously, their beaks give us a clue to their diets. The banded penguins, with their thick beaks, tend to be fish eaters. Galapagos, Humboldt and African penguins feast on the plentiful anchovies and small fish in the productive currents on the western coasts of South America and South Africa. Some of the other species of penguins that live in more temperate regions, like the Fiordland and little penguin, also forage mainly on small fish, but will also take a range of other prey such as squid and small crustaceans. Most of the species that swim in the Southern Ocean concentrate on the abundant swarms of krill that support the majority of the food web in these rough seas. The exception here is the king penguin, which eats mainly fish. Lantern fish are their favourites; these small fish are bioluminescent, which must make them easier to catch in the dark depths. Emperors catch fish too, but also eat more of a mix of krill and squid. The most diverse diet is that of the gentoo penguin, which is a very opportunistic hunter, foraging close to its breeding colonies. It will eat a wide variety of fish, squid, crustaceans and pretty much anything else it can catch.

In the early years of the 20th century, the easiest way to find out what something ate was to shoot it and cut it open to view the stomach contents directly. Strangely, when early naturalists did this with penguins they often found a collection of stones in adult penguin stomachs. At first it was thought that, as many penguins make nests out of stones, maybe they were swallowed accidentally. But as the stones were also found in king and emperor penguin stomachs, both of which do not make nests, another theory was needed; it seems that penguins must ingest them on purpose. Why they do this is still a mystery, but penguinologists have suggested a number of potential reasons, such as, as an aid to digestion, as a sort of grinding mechanism, or as ballast to help the birds dive deeper.

Luckily, today we have developed much more humane and disturbance-free methods to sample diet. We don't have to disturb the birds at all but collect the penguin's poo and chemically analyse it using a range of techniques to discover what the birds have been eating. The first scientists to do this came up with the surprising discovery that many species of penguins eat a lot of jellyfish. Although it is not their main prey, penguins often eat jellyfish, which are easy to catch but have a low energy value. As they dissolve quickly in stomach acid, evidence from stomach samples almost never revealed them.

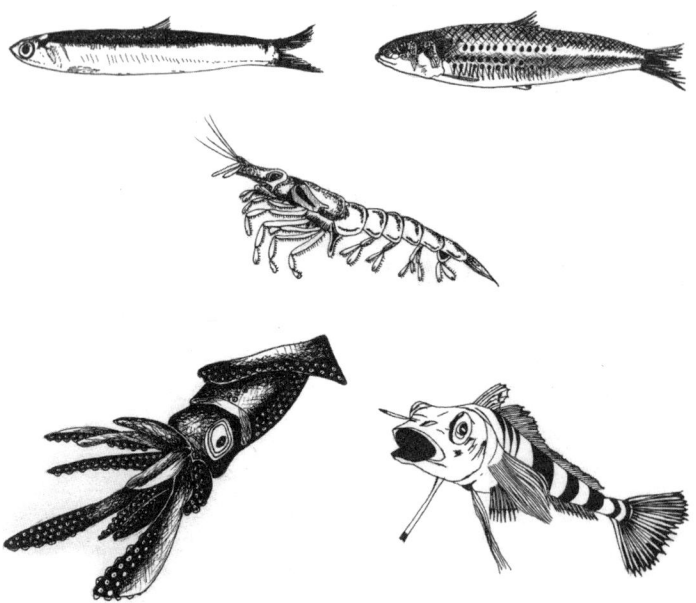

Clockwise from top left: Anchovy, sardine, ice fish, squid and krill

Predators

Lots of things like to eat penguins. Apart from their speed and agility at sea, they have few defences, and their bodies are packed with blubber and meaty muscles: a nutritious snack or main meal for any predator. At sea our little black-and-white friends are quick and nimble, out-swimming or out-manoeuvring their hunters, but on land they are clumsy and slow. They don't do too well in temperate areas where there are land predators such as foxes and big cats, although surprisingly in some regions like Southern Africa and South America they do seem to co-exist with animals such as puma, leopards and wild dogs. On land in these warmer regions, any large carnivore is a threat to an adult penguin, and smaller predators will try to eat the eggs and chicks. In Africa, crafty mongoose scurry into the penguin burrows to raid the nests, while in South America, armadillos dig into the soft earth to expose burrows and catch the helpless chicks beneath. Australia has a host of nasty critters happy to eat penguins, with Tasmanian devils, snakes and monitor lizards at the top of the list.

In many cases, to avoid these terrestrial predators, penguins and other seabirds move offshore to breed on small rocky islands where few mammals can reach them. At these island nesting sites, the main threats are airborne and come in the shape of gulls, hawks and owls. An unguarded chick will be easy prey for a hungry kelp gull, and kelp gulls are always hungry. At any of these breeding sites, the main defence against avian predators tends to be safety in numbers. Closely packed nests, with thousands of sharp beaks and vigilant parents, are the best protection against the gulls. The largest colonies seem to do better at defending themselves, with the birds in the middle of the colony, surrounded by thousands of other sharp-eyed, sharp-beaked parents, almost immune to aerial predation. A penguin nest on the edge of the colony or a couple breeding on their own is asking for trouble, and success rates for raising chicks are much better in the prime real estate in the middle of the colony, rather than at the edge.

A number of hawks and eagles, like the Galapagos hawk and striated caracara in South America and the Falklands, and the white-bellied sea eagle in Australia, will also take penguins and their young. They can often be seen gliding over the colonies, looking for stray chicks or some other easy meal to lift off and carry away.

As you go further south the number of land-based predators decreases, although the predatory birds get bigger. By the time you get to Antarctica, the only real threat to adult penguins on land are the giant petrels, or geeps as they are referred to by scientists. As their name suggests, these are the largest member of the petrel family, with a huge wingspan of up to 2 m and, unlike other smaller petrels that eat fish, these fearsome birds have developed a taste for penguins. They will take a small adult bird when out in the open but usually prefer easier pickings of chicks and juveniles if they can find them. It often takes the combined effort of mum and dad to chase them away. The geeps will play a waiting game, watching for chicks to become isolated. The highest rates of predation often occur when the juveniles leave the colony to forage for themselves for the first time. The slightly smaller skuas prey on chicks and eggs and scavenge dropped or regurgitated food. But the most disgusting bird you will find around penguins is the sheathbill. This ugly white bird actively feeds on the penguins' excrement, as well as dead chicks and eggs if it can find them. It can often be seen stalking around the edge of the colonies, and is known locally by scientists by the uncomplimentary name of the 'shit chicken'.

Despite these threats, most adult penguins are relatively safe on land. It's beneath the waves where the hunters are waiting. As penguins are so numerous, there are lots of things under the surface of the water lining up to eat them, just waiting for them to dip their toes into the sea. In warmer waters, several species of sharks, sea lions and fur seals are the main marine threats. Further south the size of penguin colonies increases and so does the number of marine predators. In Antarctica and the sub-Antarctic islands, leopard seals are the main threat. These cunning, stealthy killers will wait at the shoreline, patrolling the ice edge waiting for the birds to jump in. Out in the open water the agile penguins can often outmanoeuvre the big seals, but that is harder in the shallows when they first take the plunge. Any sizeable colony of penguins in Antarctica will have a resident leopard seal population and they often take large numbers of adult and juvenile penguins each year. When they catch a penguin they will thrash it from side to side, so that the internal flesh detaches from the skin and the seal can extract the juicy bits without having to consume the feathers – not a pleasant sight.

Leopard seals are huge and ugly, 3.4 m long with heads and jaws that remind you of a Tyrannosaurus rex. In 2006, when I first went to Antarctica, we camped near to a female leopard seal that was pupping on the beach and we watched over several weeks as the pup grew. Seeing this species raise its young is quite a rare experience, as usually these seals give birth and raise their pups out on the pack ice, well away from humans. Luckily for us, the beasts cannot move fast on land, but we still gave them a healthy distance when walking past. The mother ignored us, but the small, serpentine, evil-looking pup would slither towards us with a reptilian smile on its face. That year one of our field sites was on a small island, where a large colony of chinstrap penguins lived. The island was linked to the mainland by a 500-metre-long gravel spit. At low tide the spit was dry, but at high tide it was under about 30 cm of water. The first few times we negotiated it we happily waded through ankle-deep water in our rugged Antarctic snow boots. That was until a Chilean scientist at a nearby research station warned us that he had regularly seen leopard seals in the area surf up on to the spit to snap up unsuspecting penguins. Needless to say, after that, we waited for high tide before making the crossing.

Other Antarctic seals sometimes eat penguins but not in great numbers. The other major threat in the seas around Antarctica and the Southern Ocean, and occasionally in more temperate areas, is the killer whale. These highly intelligent, social hunters team up and deploy a number of hunting strategies to catch penguins and seals. They are one of the world's most effective hunters. There are a number of distinct types of killer whale (called ecotypes), which vary in size, colour, behaviour and diet. In Antarctica the two pack-ice 'ecotype-B' killer whales will

FLIPPERS, FEATHERS AND FEET

Giant petrel

Fur seal

Leopard seal

eat penguins. Although the larger type mostly eats seals, it will take penguins when it can, and the smaller Gerlache type specializes in our flightless feathered friends. Killer whales are massive in comparison to a penguin, so it takes a lot of penguins to feed a whole pod of them. If an orca family arrives in the neighbourhood it can decimate a small penguin population before it moves on to the next colony.

Striated caracara

Orca

Tiger shark

FLIPPERS, FEATHERS AND FEET

When they do catch a penguin they sometimes bite a small hole in the skin and suck the body out from its carcass; these whales are notoriously picky eaters and it seems they don't like feathers in their throat. I have seen many orcas patrolling along the ice edge, waiting for the penguins to jump in and planning their next meal. On one trip, a pod of about a dozen of them followed our icebreaker for three days. As the big ship cut through the pack ice, disturbing seals and penguins and making them jump off the ice into the water, the crafty killers would mop up the bewildered animals as they leaped in. These predators are huge, up to 9 m long and weighing as much as 10 tonnes, and the fin of a large male can reach nearly 2 m out of the water. It is said that they have never eaten a human in the wild, but they always send a shiver down my spine. Orcas are just that little bit too clever for my liking.

Chapter 4: BEHAVIOUR

Swimming

Penguins might be clumsy on land, but in the sea they have speed, grace and agility. Unlike most other diving birds that use their legs for propulsion in water, our birds use their stiff, blade-like wings to propel themselves. This gives them greater speed and allows deeper dives compared to any other family of birds. Over the 60 million years of their evolution, penguins have lost the ability to fly, so their body shape has changed to maximize their aptitude at sea. Their torso has become streamlined, evolving into a perfect torpedo shape to minimize drag. Their legs, which are almost redundant in the water, have shortened and moved backwards so that they are just long enough to act as a rudder, but can also be tucked back into the body so as not to disrupt that perfect torpedo shape. The stiff flippers act both as propulsion and for swift changes of direction. These stubby flippers might seem cute, but they give the birds a great advantage in the water, allowing them to accelerate, turn, brake and stop lightning-fast; they are much stronger and more dynamic than a fin. The extra agility of these birds means that the fish, squid and krill on which they prey really don't have a chance. Recent research has also shown that the unusual, scale-like feathers on penguins' wings create a sheath of bubbles which reduces underwater drag; and their flippers don't just act like a paddle, pulling the penguin through the water, but they also act like an aircraft wing, producing lift, so on the forward stroke the flipper produces thrust too. Penguins might be famous as flightless birds, but beneath the waves they really do fly.

This makes penguins fantastically successful hunters, usually out-competing other seabirds and marine predators in a local area, which allows those huge penguin colonies to form in areas where prey is in abundance. In Antarctica, scientific studies have shown that, where large emperor penguin colonies exist, the seals, Antarctica's other major predator of fish and krill, cannot compete and have to move elsewhere.

Size helps with speed, but the largest penguins are not necessarily the fastest. The speediest penguins in a straight line are gentoo penguins, which have been recorded swimming at speeds of up to 36 kilometres per hour (22 mph). This velocity can be sustained for some time and if you are on a ship in Antarctica it is quite a common occurrence to be overtaken by a group of penguins, even when the ship is steaming at full speed.

But even with these super swim speeds they are not as quick as the predators that are adapted to eat penguins. Killer whales can weigh over a thousand times more than a gentoo and are one of the fastest animals in the sea, swimming at speeds of over 55 kilometres per hour (35 mph).

In a straight line the smaller gentoo would stand no chance, but its superb agility means that it can run rings around the bigger and more ungainly predators. As long as the birds are near the shore, the penguins can use their super swimming ability to dodge and evade predators, and will often get away from a bigger opponent, but out in the open ocean, the darting penguin will eventually tire and the killer whales, which hunt in packs and have greater stamina, will wear the penguins down. For a lone penguin far from home, the chances of escape are slim, and in the end the top predators will have their supper.

But some other species such as emperors and Adélies do have an extra trick up their sleeve that makes them just as fast as a gentoo over a short distance and gives them an edge in escaping predators. They use jet propulsion. These birds trap air in between their feathers and when they need that extra boost of speed they squeeze their feathers together, pushing the air out as a jet of bubbles. Like a nitro-boost on a sports car, the sudden thrust and change in viscosity around the body of the penguin gives it an extra turn of speed, which is extremely useful if you need to escape from a hungry orca or leopard seal, or even to jump up on to the sea ice. This exit leap out of the water is often what the bubble jet is used for, as emperor and Adélie penguins need to launch from the sea on to the sea ice, the surface of which is often a metre or two above the level of the ocean. The supercharged jet of bubbles, squeezed out of their feathers, enables them to almost double their velocity to well over 35 kilometres per hour to fly out of the water by up to 3 m so that they can land safely on the ice.

Adélie leaping out of the water

The other swimming ability that helps against predators is porpoising. This is the trait of jumping out of the water in mid-swim, while travelling near the surface. Most penguins do it, especially close to the shore, where the

Gentoo porpoising

threat from predators is greatest. They will make a rhythmic series of leaps, jumping clear of the surface, often being aloft for a second or two, every 50 to 100 m. The main theory about why they make these leaps is to confuse predators, although other theories on why penguins porpoise have included saving energy and to increase their speed.

Diving

Size really does help with diving: the bigger you are, the longer you can hold your breath. Penguins are excellent divers, but they are not very big, and scientists have been fascinated by how such a small animal can dive so well. The smallest penguins, the little blues, do not dive very deep and spend most of their lives foraging within 60 m of the surface. Other small-to-medium-sized penguins dive deeper. Adélies often dive more than 100 m and have been recorded as deep as 180 m, while king penguins have been recorded at depths of over 300 m. But the daddy of the deep divers is undoubtedly the emperor penguin. The lowest deep dive recorded for the species is 564 m and adult emperors regularly dive to well over 400 m when they are foraging for food. That is incredibly deep, almost twice as far down as any other bird. It may be the biggest penguin, but diving to half a kilometre or more beneath the surface is a huge endeavour for such a small animal. At that depth the pressure is equivalent to around 50 atmospheres, enough to crush a human flat, but our super bird takes it in its stride and can do several of these deep dives every day. To perform such acts an emperor can hold its breath for over half an hour; the record length of an emperor dive is 32.7 minutes, much, much longer than any other bird. This is an incredible achievement. On average a human can hold their breath for a little over 3 minutes, and humans' lungs are much larger than theirs. Trained divers can go much deeper and the record depth for an unaided human is 131 m, but that is still less than a quarter as deep as our super feathered diver.

Scientists discovered that to get so deep for so long emperors have to take some drastic measures, and have some truly weird and slightly alien adaptations. The first adaptation, which all penguins have, is their bones. Flying birds need to minimize weight, so they have either hollow or very light bones, but underwater you need strong, sturdy bones to act against the crushing pressure, so the bones of a penguin are thick and solid. The extra weight is not a problem in the water and actually acts to reduce buoyancy. The second, stranger adaptation is their blood. The blood of many deep-diving penguin species has more red blood cells than other species. The red blood cells are the cells that transfer oxygen from the lungs to the rest of the body. The red cells in penguins have higher levels of haemoglobin, the oxygen-carrying part of the cell, which means that each individual cell can carry more oxygen. What's more, in penguins, the very make-up of the cells themselves means that they are much more efficient at transferring oxygen to the part of the body that needs it the most. This is most extreme in emperor penguins.

But it gets weirder. Even with their specially adapted blood, emperor penguins should not be able to survive for anywhere near the length of time that their long dives take. So, they have developed the ability to shut down parts of their body that they don't need when diving to minimize the use of oxygen. As their levels of blood oxygen get lower they close down the circulation to their extremities, their digestion system and their non-vital organs, so that they can just pump blood to their heart and important parts of their brain to keep themselves alive on the long trip back to the surface. On these long deep dives, the birds shut down almost their entire body, their heart rate drops from 125 beats per minute to just 5 or so beats every minute, and the birds go into a sort of stasis until they eventually reach the surface. Experiments have shown that if the blood oxygen of a mammal drops below 25% it will die, but the blood level of an emperor penguin can get down to almost zero and it will still survive and go back to do the same thing tomorrow. What a way to get your dinner!

Moving on land

Although fast, agile and graceful in the water, travelling on land is not always quite so elegant for our little comic pals. Sixty million years of evolution, to develop the ultimate swimming body, has meant that legs have become something of an afterthought. Penguins' streamlined body shape is great in the water, but on land it makes their physique rather top-heavy, with a high centre of gravity, and, with those tiny legs, things can get a bit unstable. Short legs have also given them their comic waddle. But, unlike seals and other marine predators, our birds do still need to get around out of the water, so they cannot manage without their little legs. Some species breed on steep cliffs which they need to ascend and descend between foraging trips, while other species travel many miles across land or ice each time they go out to feed their chicks. Overland travel hasn't become redundant, it's just less important.

Generally, on land penguins move in three modes: walking, hopping or tobogganing. All penguins walk, but they do so in different ways. Many of the smaller penguins can reach a fair speed, but you can't really call it running; scuttling is probably a better description. Little blue penguins are particularly good at scuttling: bending forward to lower their centre of gravity, they dart up the beach each evening on the way back to their nest sites. The beaches they run across are flat and open and often the easiest places for terrestrial predators to catch them. They need to move quickly across this open terrain before they can get to the safety of the scrub and bushes where they prefer to dig their burrows.

Most other penguins have a more upright gait than the little penguin and rely on holding out their wings for balance, although falls are still quite common. Most movement is a rather sedate waddle, an efficient and safe mode of transport, if a little ungainly and often comic to us. But if they put their minds to it, or if they are scared or angry, over short distances most penguins can muster up an impressive turn of pace, almost as fast as a human.

The only species that does not stick its flippers out for balance is the emperor penguin. The emperor will make many long, slow trips across the sea ice, from its breeding colony to the ice edge, to feed and catch food for its young. These foraging trips across the frozen wilderness can be long distances, over 50 km each way when the ice extent is at its greatest. On their way to the open water, the animals will endure wind-lashed blizzards and sub-zero temperatures, therefore the emperor holds its wings tight against its body to reduce heat loss. The emperor has no predators when it is out of the water, so it can take its time. This regal beast never runs and rarely breaks into anything more than a slow shuffle. It plods along, with a rocking motion, reserving its energy and imperial dignity across the barren wilderness of ice.

Other penguin species need to be more agile, especially those that breed on rocky islands. For most types of penguin living in the sub-Antarctic or temperate regions, the best way to avoid land predators is to go somewhere where animals with four legs cannot reach them. So, steep, remote, rocky islands are prime real estate for many smaller penguins. But getting on to steep, rocky islands requires agility, including the ability to jump up sheer inclines and between rocks, and the champion jumper, as its name suggests, is the rockhopper. Rockhoppers can jump up to 2 m in a single hop. Not bad for a creature that is only 50 cm high. That is like a human jumping four times its body length from a standing start. Most other crested penguins are good jumpers too; when you breed on rocky islands it comes with the territory. These crested penguins are brilliant climbers, with strong, sharp claws that help them cling to the slipperiest rocks and up sheer slopes. They often use their strong, sturdy beaks like a third appendage, just as a climber would use their ice axe. Their breeding sites are usually at the top of precipitous cliffs, away from the highest storm waves of the Southern Ocean, which would drench their chicks with sea spray if they were lower down. So, getting up to their nests can be a daily trial. Falls are regular; even jumping out of and into the sea can be dangerous in a storm, but penguins are hardy and can survive impacts that would kill other animals. A life in the tumultuous seas around Antarctica breeds toughness.

The penguins that live on the ice, the Adélies and emperors, have developed a third way of getting about. They toboggan. This basically means lying on your belly and pushing yourself along with those short, sturdy legs. For both of these species, tobogganing can be a more efficient way of travelling than walking. If they want to get away fast, they just flop on to their stomachs and push off. It is surprising how quickly they can move this way. Adélies will often use their wings to help, but for emperors it is

Emperors tobogganing

solely about pushing with their feet. Their strong backward-facing claws are razor sharp to grip the ice, and when the surface is slippery, they can skate along on their bellies quickly and efficiently over long distances.

The breeding cycle

Like all animals, once a penguin reaches adulthood its main focus is to reproduce. For most species of penguins this is an annual event, and this is especially so for those at higher latitudes in the Antarctic or sub-Antarctic, where the yearly cycle is dominated by the extreme seasons. They turn up at their breeding locations in the spring, going through the cycle of courtship, mating, laying, incubating, hatching and chick-rearing in few short months. There is always a rush to get the chicks ready for independent life in the ocean before the cold dark of winter kicks in and food becomes scarce. Just three species buck this trend, all of which live in more tropical or semi-tropical regions. The Galapagos penguins that live on the equator can breed at any time of year and in good years, when there is lots of food, they can have two, or even sometimes three, broods in one year. Likewise, the Humboldts, which breed off the coast of Peru, just slightly further south, often raise two broods annually. The only other species that has been recorded to have multiple broods a year is the little penguin, but most pairs have one brood. Only in a few specific places like Tasmania does this tiny penguin try for more than one brood, and then solely in the best years. For the Galapagos penguins, whose breeding locations straddle the equator, the seasons are irrelevant; what matters is the productivity of the sea, and this can vary greatly around the islands. When the cold, rich Humboldt Current is strong, its waters flow up the coast of South America as far as the Galapagos Islands, and the waters around the archipelago thrive with life. These nutrient-rich waters are exploited by plankton, which in turn is preyed upon by huge shoals of anchovies and sardines. The penguins exploit this food resource, usually successfully raising multiple broods of chicks in good years. Conversely, when the Humboldt Current is weak, its lifegiving waters do not reach the islands or the coast of South America and the ocean becomes unproductive. The fish disappear, and the adult Galapagos and Humboldt penguins have to abandon breeding, often missing entire years. In bad years starvation is common in Humboldt chicks and in very bad years the adults starve too. But when the current comes in, and the good times roll, many chicks survive each year.

The other group of penguins that have strange breeding habits are the great penguins of the *Aptenodytes* genus. Both kings and emperors have unique and very different breeding cycles, a necessity for such large birds that cannot raise a chick to adulthood in the few short months of the polar summer. The smaller penguins rely on speed to get their chicks to fledge before the bad weather sets in, but for these larger penguins, the chicks take much longer to become independent, and both species have to endure the cold harshness of the winter while breeding, but both in very different ways.

King penguins have perhaps one of the oddest and longest breeding cycles of any bird. On average, a pair will raise two chicks in three years, one chick at a time, with each cycle taking 14–16 months to complete. In the first year, the adults begin courtship in the spring, laying a single egg and incubating it on their feet throughout the summer, all in an effort to get the chick almost grown before the onset of the sub-Antarctic winter. But it is never quite enough time, so the three-quarter-grown chicks are left alone on the nest during the snow and blizzards of the winter months, only being fed by their parents occasionally, if at all. As spring returns, so do mum and dad, with food. The chicks, who may have lost half their bodyweight, once more regain their strength and complete their fledging, eventually losing their fluffy down and getting their sleek, waterproof black-and-white feathers in the spring. Getting their adult plumage is the vital step that gives them the ability to swim and become independent. The adults, in this second year, are not able to breed again until late summer when they have recuperated. In this second cycle, they will hatch their chicks just as the winter begins to bite. The younger chicks are once again left on their own on the nest over the winter period. With a smaller body and less food inside them, many do not survive. If a chick does make it in this second part of the cycle, it will take longer to fully grow and fledges in late summer. By this time, it is too late for the adults to start breeding again immediately, so they wait until the spring when they have recovered, and start the three-year rotation over again.

The emperor penguin has a less complex but, if anything, even more extreme strategy. This, the largest of the penguins, lives on the coldest place on earth, and is the only animal to breed in the Antarctic winter. Here, around the southern

King chick part moulted

coasts of the frozen continent, temperatures in winter can drop to −60°C, and cold, katabatic winds howl down from the continental interior with hurricane force. It may seem crazy to us that an animal chooses to breed in such extreme conditions, but this is the environment that the emperor is adapted for. It starts its breeding cycle in the autumn, just as the other species are leaving and going north to warmer climes. Emperors wait until the sea around the continent starts to freeze over, and then use this sea ice as a nice, flat, accessible breeding platform. Over the next two or three months, the adults go through the routine of courtship, as the temperatures drop and daylight starts to diminish. They mate in late autumn, and around mid-May the female lays a single egg, which she rests on her feet and carefully transfers to the male, who encloses it in a specially adapted brood pouch to protect it from the cold. The female then leaves the colony while the male incubates the egg. Over the next 9 weeks, during the worst of the Antarctic winter, the male carefully tends the egg, keeping it warm in his pouch and occasionally turning it. These social animals huddle in large groups often over a thousand strong to keep warm. The plan is that, when the egg hatches, sometime in late July or early August, the female will turn up with a belly full of food. If the wife is late, the male emperor has a special gland in the back of his throat that can excrete a waxy substance that can keep the chick alive until she returns.

Once mum returns, the exhausted and starving male can go back out to forage in the ocean and regain his strength, before coming back once more to feed the chick. Then the race is on to get enough food to raise the youngster. At first, the parents take turns to feed the chick. One will keep the youngster warm in its pouch until the other returns, but by October the chick is often too big for the pouch and large enough to fend for itself against the cold winds and the

Emperor chicks huddling together to keep warm

hungry skuas. So, then both adults go out to sea to find food, while the chicks stay together in crèches, sometimes huddling together when the weather turns bad.

If all goes well and the ice remains stable, the adults will give the chicks their last meal in December, just as the chicks are starting to transform into the sleek watertight adult plumage. As they become waterproof, pangs of hunger will drive them to leave the colony and enter the cold waters to fend for themselves.

Fighting

All penguins are social creatures that breed together in huge, densely packed colonies, rubbing shoulders with tens of thousands of neighbours. Here, in these avian metropolises, they compete for mating rights and nest sites, and travel through the colony commuting across, or close to, other birds' territories. Within a colony, there is always a lot of bickering, aggression and fighting. They may seem sweet and fluffy to us, but if you are a penguin, you should annoy a fellow penguin at your peril. Fights are common and can result in serious injury. And fighting is not only for the males; in several species the girls fight just as much as, and sometimes even more than, the boys. But before it gets to all-out war, these clever birds have developed complex displays of aggressive and submissive behaviour that usually avert a battle. Threat behaviours are common and are often different for each type or family of penguin. Some of the larger species, such as king penguins, crane their necks, point their bills towards the target and circle their heads around, like a fencer circling their wrist, before they jab their foil at the opponent.

BEHAVIOUR

King threat posture

Others, such as the Adélie, rub their bill on their wing, a little like sharpening their blade before striking. A common aggressive posture for many of the smaller penguins is the 'sideways stare', pointing the beak away from the target and fixing their stare at their opponent with a single eye. In the case of crested penguins, this can be exaggerated by them raising their crest or, in the case of the Adélie, raising the feathers on the back of their heads. Adélies also have the ability to widen their eyes, exposing the white rim around the eye socket to make their eyes look bigger and scarier.

The banded penguins often use an 'alternate stare'

posture, where the bird bends and looks directly at the opponent, then rotates their head to look first through one eye, then rotates the other way to stare through the other. If you bump into a banded penguin looking out of its burrow, you will often be met with this strange gesture, as the bird rapidly rotates its head looking at you with alternate eyes. Other aggressive actions include growling, head bobbing, flipper flapping and hissing.

The target of the aggression usually has two options: they can act submissively or front it up. Submissive behaviour often involves a calculated withdrawal. Penguins have some quite comical walks associated with this. It is a common strategy when birds are commuting through a densely packed nesting area, trying to avoid getting too close to other nesting birds on their way to and from their own patch. Little penguins adopt a 'low walk' posture where the bird arches its back and keeps its bill parallel to the ground, scuttling away from the aggressor as fast as it can.

Little penguin, low walk

King penguins and several other species adopt a 'slender walk' posture when navigating through the crowd. They raise their frame almost on to tiptoes, keeping their torso as tall and thin as possible, with their head held high and bill pointing downwards, walking as carefully as possible between all those sharp pointy beaks. On the nest, if your neighbour starts to kick off, dismissive rather than submissive behaviour is often the key. This can involve rolling the eyes and turning the other way, or pretend pruning, in an effort to ignore the aggressor.

Often these actions work, but sometimes neither penguin is ready to back down. At this stage there is frequently some

Kings' slender walk

Macaronis fighting

sparring before all-out attack. Wing flapping and long-range flipper fencing can ensue. In crested penguins and gentoos, bill jousting is common. This entails birds interlocking their bills and rapidly twisting their heads in an effort to flip the opponent over. When all else fails, the birds engage full on with flipper bashing, pecking and biting each other until one antagonist is injured or decides to make a run for it, often chased by their aggressor. In these full-on fights injuries are common and can be serious. Fighting is most common during courtship and just before egg laying. During incubation it often dies down, only to resume when the chicks hatch and there is a lot of movement of adults through colonies, going out or returning from foraging and upsetting the applecart. Some species are less aggressive than others. Generally, penguins that live in the largest, most packed breeding sites and defend their nests tend to fight more, with Adélies, chinstraps and rockhoppers most inclined to aggression.

The burrowing penguins tend to nest at lower densities, so are less crowded, whereas the brush-tailed species sit on nests that are pecking distance apart. The most serene and less inclined to aggression is the emperor, which has no nest and rarely fights. Indeed, in the winter this imperious bird must huddle together with thousands of its counterparts to survive, so compromise is the order of the day, and the only thing to fight against is the fierce Antarctic weather.

Courtship and mating

Penguins are great romantics. They have elaborate courtship routines and show great devotion to each other, and many species will pair for life, or at least for many years.

Almost all penguins breed in colonies, which usually number from a few hundred to tens, sometimes hundreds, of thousands. At the start of the breeding season, the males will arrive at the breeding site a week or two before the females to reserve a nest site and practise their courtship routines. Often they will return to the same nest each year. Once the females arrive, the action really starts. Males bray and trumpet, advertising their wares to prospective partners. Head back, beak in the air, the male lets rip at the top of his voice. Unlike other more tuneful birds, for a penguin it seems to be mostly about volume rather than melody. It's like every ounce of their body goes into the call, a routine that scientists call an 'ecstatic display'. In some species the males have additional ways of advertising their availability: some will flap, and others pose with their wings either outstretched or close to their bodies. Yellow-eyed penguins throb and shake, pulsating the feathers of their neck and throat, making a chuckling call. Male king penguins often stand very still with their flippers down and beak raised to the heavens, while the crested penguins bob their heads and raise their crests to try to attract a mate. In some species, the females also join in the party. The females of African penguins will call to attract the boys; not surprising really, as the male African penguin has probably the most complex and varied set of courtship moves, which include flapping, head rotating, vibrating, bobbing and the usual ecstatic trumpeting.

Once a female shows interest, the display changes and both sexes join in. In some species this mutual display is similar to the ecstatic braying of the lone male, but now the female also starts to bray and trumpet. But in many species the show is more complex. Chinstraps display alternative trumpeting, with one partner raising their head skywards and giving their all while the other bends down towards the ground, and then the roles are reversed and the other partner sings while the first bows down. For macaroni penguins, this develops into a dance-like display, with additional mutual head swinging, wing flapping and calling.

After this initial test of finding a potential mate is complete, the next stage is to show devotion, to cement the pair-bond and prove their commitment to each other. Each species tends to do this in a different way. Yellow-eyed penguins will walk around the nest in a tight circle, while a pair of emperors, which have no nest, will walk together side by side through the colony, each showing off, while they promenade with their respective partner.

Many species will bow their heads down towards each other in unison. These bobbing movements can be quite simple, but in some species they have become more complex, with quivering and growling or hissing. One species, the African penguin, has taken courtship to another level with an extensive repertoire of moves, including what scientists name the 'extreme bow', the 'oblique stare bow' and 'mutual beak slapping'. This exuberant bird really is the John Travolta of the penguin world. Once the pair-bond is stable, they will demonstrate other ceremonial activities such as collecting stones and intense nest-building.

The final act is that of mating itself. This is a fairly quick affair, but will happen multiple times between pairs over the course of a few days, often several times an hour. The male approaches the female, leaning on her back, and if she is willing, she will submit to a prone position and the male will then hop on to her back. There is often a lot of neck preening, nibbling and bill caressing, or in some species the male will flap noisily, vibrate his wings or press the female's

Chinstraps copulating

flippers down. For most species it doesn't last long and, as soon as it is over, both penguins jump up and preen themselves.

Nesting

Almost all penguins build nests. These can be complex, permanent structures or simple scrapes in the soil, with a bit of vegetation for decoration. Most nest-building and breeding happens at a specific site where lots of penguins come together each year to find a mate and raise their young. In the penguin world these are called colonies, or rookeries. These penguin cities can number hundreds of thousands of densely packed nests, stretching for miles, and they can take over entire islands. This home-building can significantly change the landscape. The size of colonies depends on the species; yellow-eyed and Fiordland penguins are less social and hardly have any colonies at all, gentoo colonies usually number just a few hundred, but kings, chinstrap, royal and macaroni penguins can number many hundreds of thousands at a single site. The world's largest penguin colony is thought to be on Zavodovski Island in the South Sandwich archipelago. This small island is an active volcano, just a few kilometres across. It hosts over 1 million penguins, a mix of mostly chinstrap and macaroni. With many active volcanic fumaroles belching sulphurous fumes, and over 1 million penguins, Zavodovski Island is often given the title of the smelliest place on earth.

Colonies with a mix of species are quite common, especially in Antarctica and the Southern Ocean. At some sites you can see all three species of brush-tailed penguin living side by side with one another. These three birds prefer slightly different nesting places, and mostly it depends on who gets to the nesting place first. The Adélies arrive at the colony earliest and bag the prime real estate on the crests of the hills, the gentoos come next and go for the ridgelines, and finally the chinstraps arrive and have to put up with the slopes. If they get there too late they might have to breed on the edge and be a target for the skuas, or on the flat bottoms of the colony, which turn into drainage ditches when the snow melts.

It helps to build your nest as close as you can to your neighbour, as many beaks can deter aerial predators like skuas and gulls. But it is best not to build too close, otherwise they will be within pecking distance. Penguins also defecate out of the nest, so building an abode too close could be a messy affair. Overall, nest patterns tend to self-organize, pecking or pooing distance apart. Depending on the size of the penguin, the nesting density in a colony can be up to 2.4 nests per square metre, with the rockhopper having the densest colonies. Studies have shown that the most mature or aggressive penguins get the best nests in the middle of the colony, and usually have the best breeding success.

Building a nest can be a long process. Each season it is started by the male before the female arrives and construction continues throughout courtship and incubation, one stone at a time. A good nest can help attract a female, so males put a lot of effort into their build. Once paired up, the penguins will often try to use the same nests year after year, so it's useful to have a good sturdy construction to avoid the eggs rolling away or being flooded. But they have to be careful, as stone stealing is big business. While you are out industriously collecting the best stones for your pad, your next-door neighbour might be back at the nest nicking all your best building material. It has also been noted that some females will sell themselves for a good stone!

Adélie nest-building with rocks

Over the years, nests grow and can become quite large. One scientist once counted the number of pebbles in a medium-sized gentoo nest and was surprised to find that there were over 1,700 stones. But it makes sense to have a big, sturdy nest, as storms in Antarctica and the Southern Ocean can be tempestuous and these penguin cities last for thousands of years. Radiocarbon dating has proven that some colonies and nests have been in almost constant use since the glacial ice retreated from the beaches at the end of the last ice age, 8,000 years ago.

At some sites that I have visited on the South Shetland Islands, the penguins have built their colonies on a series of beach ridges, stretching back a few hundred metres or so from the sea. Over time, the penguins have taken stones from close by and added them to the location of the nests, lowering the height of the beach ridge outside the colony and raising it around the nests, so each ridge crest, which was originally level, has become a series of symmetrical, circular mounds, creating a checkerboard-like pattern in the landscape. So, not only do the penguins build cities, they are the architects of the landscape around them too.

But there are other types of nests. Banded penguins live in semi-tropical environments and need to keep out of the sun, so they tend to build burrows or nests in caves and hollows. Their burrows are simple affairs, with metre-long tunnels and a small chamber at the back, with a little bit of dead vegetation for comfort. Fiordland, Snares and yellow-eyed penguins breed in forested areas, often in shrubby vegetated slopes around the coast. These penguins breed at the lowest densities, where each nest may be several metres apart. For these species, deforestation has become a problem over the last century when humans have cleared the forests for farming or development.

The only two species of penguin that don't build nests are the great penguins, the kings and emperors. These species lay a single egg and carry it on their feet. They both have a special flap of skin that covers the egg to keep it warm called a brood pouch, just big enough to house the egg, or a young chick, and keep it away from the worst ravages of the weather. Kings will at least stay in one place and keep a short distance apart from their neighbours while incubating the eggs, but emperors are always on the move and can shuffle several kilometres with their precious cargo on their feet.

Eggs and incubation

All penguin eggs are pretty much white or off-white, no matter what the species. The eggs themselves are more roundish than those of most other birds and have thick shells, probably because most penguin nests are often made of rocks, so they need to be tough. Interestingly, compared to a penguin's weight, their eggs are some of the smallest of any bird and are proportionally smaller the larger the species. Emperors are thought to have one of the smallest egg-to-adult mass ratios of any avian species. The great penguins, the kings and emperors, only ever lay one egg per season, but the smaller species lay two eggs per clutch, although occasionally some youngsters will only lay a single egg, but it's fairly rare. Several species will develop spare eggs in the womb, so that if an egg breaks early in the season, they can quickly lay another one, but you will almost never see three eggs in a nest.

The banded and brush-tailed penguins lay eggs that are about the same size, but the crested penguins have a strange adaptation of laying odd-sized eggs, with the first significantly smaller than the second. The Magellanic penguin always lays two eggs; the first is small, only 60% of the size of the second, which is laid a few days later. Scientists believe that this unusual strategy is so that one chick always has a high chance of survival. When food is scarce, the first chick often dies and only the larger second chick will survive. The adults will feed the strongest chick first, to ensure it survives, rather than both chicks dying. In a good year, both chicks will usually survive.

Incubation of the eggs takes a few weeks, less for the

smaller species and longer for the larger penguins. The little penguins' chicks will hatch within 5 weeks, the shortest hatching period, but for the emperor it takes around 8 to 10 weeks. During this time the egg is always tended, guarded or, in the case of the great penguins, held in the brood pouch on top of the adults' feet. For all species except the emperor, both mum and dad take turns to look after the egg, but for the emperor this is solely a male endeavour, carrying and protecting the egg throughout the long incubation period in the cold and dark of the Antarctic winter. The hatching of the egg, through that tough shell, is a lengthy process and can take between 20 hours and several days for the larger penguins.

Gentoo with two eggs

Chicks

When the chicks hatch, they are usually small and scrawny, not the cute, fluffy bundles that we normally associate with penguin chicks. At this stage, they need to be brooded continually by the parents. The skinny youngsters cannot initially maintain their body temperature, so it is essential that they are kept warm and dry for the first few days and weeks. The adults will rotate the care duty, with one parent on the nest and the other catching food out at sea for the hungry offspring, swapping roles when they return. Researchers term this phase of chick-rearing the guard phase. After a few weeks (or longer for yellow-eyed, kings and emperors) the chicks will become more independent, being able to regulate their own temperature and now big enough to fend off smaller predators. They will also have grown too large to fit under their parents' bodies or in the brood pouch, and will start to take their first steps away from the nest, although at this stage a sudden scare will send them scurrying back to the nest to seek the shelter of mum or dad. It can make a funny sight, seeing an overgrown chick running back to its parents and trying to force itself into the brood pouch, which is way too small for it; often only the head will fit in, leaving its rear and legs scrabbling outside.

At this stage the chicks reach maximum cuteness. As a scientist I am not really supposed to utter words like 'cute' or 'adorable', but if you ever get to see an emperor penguin chick in the wild, they really are something special. I have seen hardened biologists go all dewy-eyed and wobbly when seeing their very first emperor chicks. That loveable panda

Emperor penguin chick and friend, pushing into the brood pouch

face, their fluffy grey down and their comic character get me every time.

Not that every species of penguin has pretty or handsome offspring, and propping up the bottom of that list is the chick of the king penguin. Even though it is the closest relative of the emperor, its chicks could not be more different. For a start they are a rather dirty, undistinguished

King chick

dark brown in colour. You couldn't call them ugly, but they are not much of a looker. Unlike their smart and fashion-conscious parents, which are considered the most elegant and well-dressed of any penguin, the chicks are drab, dull and untidy. They are often also very large, as they have to spend the whole winter on the nest, mostly without food, so they try to grow to almost the size and weight of their parents before they are left to tough out the storms of the Southern Ocean. They overwinter at their breeding sites in their thick layer of brown, downy feathers. It is said that when early explorers first visited the Southern Ocean islands and found the king penguin chicks, often on their own, in large numbers, with no sign of the adults, they looked so different that they thought that these birds were a completely different species. This belief persisted well into the 19th century. Field assistants and scientists who work with the chicks today still call them 'woolly bears', as the youngsters stoically sit there, waiting for the spring and the return of their parents.

Once a chick gains some independence, its parents will stop guarding it and mum and dad will both go into the ocean in search of more food. In many species, the chicks group together to form crèches. These groups help the youngsters to fend off predators and, in colder climates, keep the young warm when the weather turns bad. For emperor penguins, the youngsters will huddle in large groups like their parents, to survive the freezing winds and blizzards that often occur around the Antarctic coastline. On better days, the chicks may wander around the colony as, now that the adults have left and the nests are empty, there is less territorial aggression. But the chicks of most nesting species don't wander too far from the nest. However, when they do, this can create a problem when the adults return to feed their chicks, as the youngsters are often not where mum or dad left them. Working out which chick is yours, when there are thousands of them around and they all look pretty much the same, can be tricky. Adult penguins have two ways of identifying their own offspring. The first is by voice. For many species, the call of each individual is subtly unique, which helps the parent identify its young. A penguin colony at this late stage in the season can be a very noisy place, with youngsters continually calling, and parents answering back to try to find each other. The second way is a slightly strange behaviour involving the 'chick chase'. Voices can be impersonated and even when the parent thinks that it has found its chick using its unique call, there will often be other hungry chicks, trying to steal a meal or generally barge in and get fed. Chicks at this time can be quite large and imposters could disrupt feeding and waste that hard-earnt fish. To get away from the pretenders, the parents adopt a rather strange strategy; they run away. The chicks will often follow as close as they can – travelling through the colony, where an aggressive peck from a neighbour is never too far away. These chases can last for several minutes and cover some distance, but generally the real offspring will be the only ones that persist with the chase and all the others will give up earlier in the running, so eventually the adults will be able to single out their own chicks from the crowd.

The last stage for the chick before adulthood is moulting, when the young replace their fluffy juvenile feathers with sleek black-and-white plumage. Until this time the chicks are not waterproof and cannot go into the water to feed for themselves. The downy fluff is good for keeping out the cold and wind, but it is useless at repelling water. Adult feathers are very different from the downy fluff of the young and provide warmth and a water-resistant barrier against the freezing ice-filled waters around Antarctica. Towards the end of the chick's development, their feathers start to change, as the longer, stiff adult feathers push the old fluffy ones out, a process that can take several weeks until the chicks metamorphosize into adult plumage. During this time the birds can look quite untidy and bedraggled as they shed their downy coating. At this stage, the adults will leave the breeding site, not to return until next year. Chicks will follow when they are able, sooner in the more northerly species, which do not have to brave the colder waters, but more often several weeks after the parents for emperors, when pangs of hunger drive the juveniles into the water to find food.

Moulting and fasting

It is not only the chicks that moult. There is one last thing that an adult penguin needs to do each year after breeding and that is renew its feathers. Generally, all bird species replace their feathers annually, a process called moulting, where the old, worn or damaged feathers drop off and new ones take their place. Other families of birds tend to do this in a piecemeal fashion throughout the year, but penguins change their feathery coat all at once over a few weeks after the breeding season is finished, in a process that scientists call a 'catastrophic moult'. This isn't to say that our little friends lose all their plumage and are as bald as a plucked chicken before they get a new coat: that would not be a clever strategy in the cold climates of the Southern Ocean! For penguins, as for most birds, new feathers start to grow underneath the old ones and push the old quills out, so a penguin is never naked and always has some protection against the cold. For macaroni penguins, the new feathers are already half grown before the old ones drop off, although for other species the new feathers tend to be only a quarter to a third of the original length as the older ones are discarded.

Almost all adult penguins moult in the summer, just after they have left their chicks to fend on their own, although some of the crested penguins, which live in more temperate climes, will moult in the autumn. Most species will return to their colonies and sit on a nest or inhabit a burrow while the process occurs, although the two true Antarctic species, the Adélies and emperors, will find stable patches of sea ice to sit out their moult. While they are shedding feathers, penguins look very untidy and bedraggled, a rather sad affair, made worse by the fact that at this time they are not waterproof and cannot go out to forage. So, all they can do is sit quietly and wait for the process to unfold. Moulting typically takes a few weeks, but is quicker for the smaller penguins and longer for kings and emperors. A Galapagos penguin may replace its plumage in as little as 13 days, but for an emperor, with all those feathers, it might take 5 weeks. The Galapagos penguin is a bit of an exception to the pattern of penguin moulting, as it usually moults before breeding and often twice a year, probably because it lives on the equator and the seasons are somewhat different at those latitudes.

The biological process of growing new feathers takes a lot of energy, and as the penguins cannot eat or replenish their reserves, they tend to lose a lot of weight during this period. Macaroni penguins, which have a short and energetic moulting process, can lose up to 50% of their bodyweight while renewing their feathers. Usually this has been compensated for beforehand as the birds will have been frenetically fattening up ready for this enforced period of weight loss.

It is not the only time when penguins go through a fasting process. When they first arrive at their colonies and start the courtship and mating process, most penguins will endure several weeks of enforced fasting. This can last between a few weeks and several months. A female macaroni may spend 42 days onshore, between when it first arrives at its nest and when its egg hatches and it can go back to feed. During this time it may lose 40% of its weight, but this pales into insignificance when you consider that a male emperor will spend 115 days without food in the Antarctic winter, huddling in the freezing winds, often losing over 50% of its body mass to protect its precious egg. Even more extreme is the process for king penguin chicks, which will spend all winter, up to 5 months, sitting alone at the breeding site waiting for their parents to come home to feed them. Researchers found that, on average, each chick lost about half of its bodyweight before being relieved. An astonishing breeding strategy.

So, in the life of a penguin, as well as extreme endurance, romantic courtship and a frenetic breeding cycle, we can add crash diets into their extraordinary lifestyle.

Vocalization

Anyone who has ever visited a penguin colony will tell you that they are very noisy animals. They will also tell you that they are not the most musical of birds. Penguin calls are loud, with words like 'raucous', 'grating' and 'harsh' often used to describe the sound they make. As many colonies will have thousands of penguins all shouting at once, the noise at a colony can be a real racket. Most calls are loud, flat and from a low starting note rise in pitch for several seconds, a little like what I imagine trying to start a rusty chainsaw might sound like. If you are unfortunate enough to camp near a penguin colony in Antarctica, where the sun never sets, the penguins never quieten down. They call throughout the day and night, so sleep can be difficult to come by, even if you can acclimatize to the stench.

Penguins vocalize for a number of reasons, including courtship, squabbling, communicating or trying to find

Magellanic penguin calling

their chicks. This does mean that for some parts of the breeding season, such as during incubation, colonies can be a little less noisy, until the chicks hatch and add to the cacophony. Before that, in the courtship phase, the male does the asking, as he advertises his wares to the opposite sex by raising his head skywards and giving his everything, trumpeting or braying to the heavens for all he's worth.

This courtship call, which for most species usually includes flipper waving or head wagging, is termed 'an ecstatic display' by researchers. All penguins except the emperor call this way. Our imperial friend usually trumpets with his head down, rather than pointing heavenwards, and tends to flap much less than his smaller cousins, probably to conserve heat. Although their calls seem far from musical to the human ear, to a penguin they are full of meaning. To the discerning female, a courting call can identify between a big, strong, healthy male with a loud, deep, booming voice and a young, skinny pretender with a higher-pitched, weaker call.

Later, when the adult needs to find their chick, a different call is used to identify their own offspring. All penguins will call to their young, especially later in the rearing season when the chicks are more mobile. Some penguins, such as the little blue penguin, have a large range of different calls for many social situations. These include different grunts, beeps, brays and roars for interactions such as courtship, aggression, defensive behaviour, greeting their partner or pair-bonding.

Emperor penguins each have complex and totally unique songs that use ultrasound, unperceivable to the human ear, to call to their chicks. The adult emperor's call is the most complex of any penguin. It needs to be; emperors don't have nests, and towards the end of the breeding season the chicks can be very mobile. When the adult returns to feed its chick in November or early December, the colony will have split into a number of sub-colonies or suburbs, usually with several hundred chicks in each. These small groups probably form so that the chicks are easier to find, and their calls are not drowned out by thousands of other voices that would ensue if the whole colony came together. The sub-colonies are quite mobile and might have moved many hundreds of metres in the 4 or 5 days since mum or dad went out to get food. As all the chicks look pretty much identical, the only way to tell them apart is by their voices. None of those ridiculous chick chases for the regal emperor; they, as previously mentioned, are far too sedate and sensible for such nonsense, so they rely solely on their unique call. The emperor chicks are constantly calling, and their voice is a pleasant high-pitched fluting sound. If you listen carefully, it is clear that each is slightly different. The sound of an emperor colony late in the season can be quite wonderful, with the noise of thousands of chicks singing their high-pitched melodies, interspersed with the occasional deep baritone rumble of an adult trumpeting, as they home in on their offspring.

Pengwin (Welsh)

Penguin (English)

Pinguin (German)

Manchot (French)

Pinguino (Italian)

ペンギン Pengin (Japanese)

पेंगुइन Penguin (Hindi)

Pingvin (Norwegian)

Pingüino (Spanish)

企鹅 Qi'é (Chinese)

Korara (Māori)

Chapter 5:
PENGUINS AND PEOPLE

Discovery

It's not entirely clear who discovered the first penguins, although in this context 'discovery' is a very Eurocentric term. In reality, the native people of South America, Africa and Australia have evolved with these birds over hundreds and thousands of years. Saying that Western cultures discovered them is a rather old-fashioned viewpoint. From a European perspective, the first explorers to encounter penguins were probably aboard Bartolomeu Dias' 1488 expedition to the Cape of Good Hope in South Africa, but although there are several penguin colonies in the area, no records of strange black-and-white birds came back from that trip. Álvaro Velho, a sailor on another Portuguese expedition, the voyage of Vasco da Gama around the Cape to India in 1497, published a travel log of the journey where he recounted finding birds *'as big as ducks, but can't fly because they have no feathers on their wings. These birds, of which we slaughtered as many as we could, cried like donkeys.'* The birds were almost certainly African penguins, which still live around the Cape today. This, in all probability, is the first written record of any penguin.

A few years later, in 1520, when Ferdinand Magellan sailed around South America, he too encountered penguins. The chronicler of his travels, Antonio Pigafetta, documented, *'we found two islands full of geese and sea lions, but the geese were so many that it was impossible to count them; we filled the five ships with them for an hour. These geese are black, and they have feathers all over the body of the same size and shape. They do not fly and feed on fish, and they are so fat that they were difficult to pluck, but we took off their skin. They have beaks like those of a raven.'* These poor creatures would have been Magellanic penguins, named after the famous explorer leading their expedition.

The first reference to the term 'penguin' dates to 1577, in the logbook of Francis Drake's ship, the *Golden Hind*, as he battled around the southern tip of South America. The account, written by one Francis Fletcher, a priest on board, records, *'In these Islands we found great reliefe and plenty of good victualls, for infinite were the number of fowle, which the Welsh men named Pengwin . . . [The birds] breed and lodge at land, and in the day tyme goe downe to the sea to feed, being soe fatt that they can but goe, and their skins cannot be taken from their bodyes without tearing off the flesh, because of their exceeding fatnes.'*

The comment on the Welsh name, 'Pengwin', is probably a reference to another bird, a resident of the northern hemisphere rather than the Southern Ocean. It seems that Fletcher was mistaking the Magellanic penguins for great auks, a very similar, but totally unrelated, black-and-white flightless seabird that lived at the time on the North Atlantic coasts of North America, Greenland, Iceland and Britain. This bird grew up to 80 cm tall and was sadly hunted to extinction by European seafarers in the mid-19th century.

Their Latin name, *Pinguinus impennis,* probably comes from the Welsh words, *pen gwyn* meaning, 'white head', and it seems likely that this is the name that Fletcher was referring to. Unlike Magellanic penguins, which don't have white heads, great auks have a prominent white patch above their beaks – hence the Welsh name. So Fletcher was wrong on two counts: penguins were not auks and they did not have white heads. But the name stuck and over time seems to have morphed from 'Pengwin' to 'penguin' and is now universally accepted.

There are other theories on how penguins got their name. One story relates to a Dutch expedition in 1598, also to the Straits of Magellan, calling the birds 'penguins' due to the Latin *pinguis*, meaning 'fat' or 'oily', but it seems that this reference comes later and is slightly less believable than Fletcher's logbook.

The Latin name for the penguin family is *Spheniscidae*. The word comes from the Greek word for 'wedge', used to describe the shape of an African penguin's wings, which have evolved into swimming flippers. Interestingly, they were given the name in 1831 by an eminent French ornithologist, Charles Lucien Bonaparte, the nephew of Napoleon. So penguins really do have an imperial heritage.

Exploitation

People have always exploited penguins. Where native people lived and evolved over time close to penguin colonies, they hunted them for meat, skins and eggs. We know that the native peoples of South America often used this local resource for both food and clothing. There is evidence of the aboriginal people of southeast Australia hunting them and of the native people around the Cape in South Africa collecting their eggs. In these places, where humans and penguins have existed together for millennia, exploitation, as far as we know, has not impacted the long-term populations. Nature has co-existed with humans. But in areas where new settlers have moved in and discovered penguins, the results have often been disastrous.

It started some centuries ago. The first and, luckily, so far the only recorded extinction of a penguin species due to over-exploitation occurred when Polynesian settlers colonized the Chatham Islands. This group of small remote islands, in the Pacific east of New Zealand, had a number of indigenous endemic birds and animals living on them. The Chatham Island penguin was thought to be a distinct species, closely related to other crested penguin types in the region. Like many island animals, they had evolved over time in isolation, diverging from their closest relatives, restricted to a few small islands; they were never numerous. The Polynesians moved into the Chatham Islands around 1500 AD, and within 200 years the penguins were gone. The penguins went the same way as the dodo in Mauritius, the rhea in New Zealand and many other unique endemic species that are restricted to island homes.

When Europeans first encountered penguins on their voyages of discovery in the age of exploration, they were seen as an excellent source of food and resupply. To the sailors and explorers braving the challenges of long sea journeys around South America and Southern Africa, suffering with scurvy and terrible diets, the bird colonies must have seemed like a godsend. Penguins were small, slow on land and had very little fear of humans. What's more, their colonies were huge, often thousands or tens of thousands strong. A seemingly unlimited supply of fresh meat. Francis Drake recorded killing 3,000 birds near the Strait of Magellan and it seems unlikely that his circumnavigation would have succeeded without this supply of fresh victuals. Although cruel, and devastating at the time, these early infrequent visits probably did not harm the long-term populations of the penguin colonies around Tierra del Fuego.

Likewise, more recent explorers from the golden age of polar exploration have often resorted to penguins as a ready food supply in a crisis. Several polar expeditions, like those of Larsen or Borchgrevink, who both inadvertently overwintered in different parts of the continent, survived on penguin meat. Not that it was always appreciated. Tastes differ. Some records suggest that penguin tastes like goose, but many others complain of its fishy flavour. The eggs were no better. Eggs tasting of fish were not everyone's cup of tea. Especially as in the Antarctic penguin eggs have special proteins, adapted to the cold, that make the whites remain translucent and rubbery after cooking and turn the yokes a blood-red colour. Needless to say, seal meat was often the preferred choice over penguin, even for the starving polar explorer.

As seafaring voyages around the Cape of Good Hope from Europe to the East Indies became more frequent, penguin eggs became a useful source of fresh food for many sailors. At the height of this marine trade route near the end of the 19th century, up to 700,000 penguin eggs a year were being collected. Packed into crates and surrounded by soft sand to keep them cool, the eggs could last several weeks and must have been a welcome change to maggot-eaten sea biscuits. Over a thirty-year period around 13 million African penguin eggs were collected, leading to the total loss of many colonies. In some places, such as the Falkland Islands, penguin eggs became an important source of protein for local people and collecting penguin eggs for food continued until almost the end of the 20th century.

Egg collecting on the Falklands

It was not the only thing that the people of the Falklands and South America used penguins for. The skins had always been used by the native tribes for clothes, and the cities of South America had a thriving trade in pelts and feathers. Humboldt and Magellanic penguins that bred in vast numbers around the southern tip of South America seem to have been preferred, with skins being transformed into hats, purses and muffs and the feathers used to trim women's dresses, and to decorate their hats.

Commercial exploitation of Southern Ocean penguins on an industrial scale did not really start until the late 19th and early 20th centuries. The worst case was not for their skins, eggs or flesh, but for their oil. Of course, by then, exploitation of the many bounties of the Southern Ocean was well under way. The sealers had almost annihilated the fur and elephant seal populations around most of the Southern Ocean islands, while the whalers were busy fishing out all the whales from the depths. Penguins were not exempt either and where they existed close to sealing and whaling stations, many hundreds and probably thousands went into the sealing trypots to be boiled down for their oil.

But a single small penguin didn't produce enough oil to make their mass slaughter profitable on an individual basis, so the early hunters focused on larger prey. That changed in the 1890s, when an enterprising and ruthless New Zealander called Joseph Hatch invented the penguin digester.

Hatch was raised in New Zealand and was a wealthy and leading local businessman, who was elected Mayor of Invercargill in 1877. He went on to become an elected member of the New Zealand parliament in 1884. He had made his fortune in the lucrative business of sealing, but his political reputation was shattered when it was discovered that his sealing ships had been hunting seals out of season, defying the laws set by his own government. The scandal led to him losing his seat, and ended his political career. So, he moved his business to Tasmania, refocusing his seal hunting on the Australian Southern Ocean Territory of Macquarie Island.

This uninhabited, storm-lashed island, 1,000 km south of New Zealand, was a haven for wildlife. Hatch had first set eyes on it years before and had commented on the multitudes of seals and penguins. After his move to Tasmania, he sent his sealing ships to this remote and, up until then, pristine outpost. Initially, he concentrated his efforts on catching elephant seals, boiling them down for their oil. The process was grizzly and simple: kill the large elephant seal bulls, which could each weigh up to 5 tonnes, cut off their blubber and melt it down in large cast-iron cauldrons called trypots. There were thousands of elephant seals on Macquarie and the first few years were lucrative if dangerous work: Hatch was quite literally making a killing. But as the seals became fewer in number and the haul-outs closest to the trypots were depleted, production began to dwindle.

This was when Hatch had his eureka moment, an idea that propelled him from a run-of-the-mill sealer to one of the most infamous entrepreneurs of his day. During his travels he had seen a recent invention in the whaling industry, a huge, pressurized canister that could boil down whales to extract oil not just from the blubber but from the bones, skin and entire carcass of the animal. Hatch realized that with a few adjustments this diabolical mechanism could be converted for processing penguins. As he did not have to skin each penguin and could process them hundreds at a time they suddenly became a viable business, and so, with a few tweaks to the original boiler, the penguin digester came into being. The digesters were installed right in the middle of penguin colonies. Each would be filled up to the brim with dead penguins, a single digester taking up to 1,000 penguins at a time, and after boiling them for 3 days, each penguin carcass would produce 1 pint (0.5 litres) of high-quality machine oil. His aim was *'a pint per penguin'*. King penguins were targeted first, but after these were all but wiped out, larger digesters were installed at the colonies of the smaller royal penguin, a penguin that only bred on Macquarie Island, where it lived in vast colonies of almost innumerable size.

By the turn of the 20th century the outcry against the penguins being slaughtered started to mount. Very few people ever visited the remote Macquarie Island, but it was conveniently situated on the way back from many heroic Antarctic expeditions at the time, and several high-profile explorers saw and commented on the penguin butchery. The first was Edward Wilson, the naturalist from Captain Scott's first expedition in 1901, who, on his return to Britain, stirred up public sentiment against the Macquarie operation. But sentiment and letters of complaint were a poor return against the profit Hatch was making from penguin oil. The entrepreneur had invested big, increasing the number of digesters, and the profits were rolling in. Tasmania, and the city of Hobart where Hatch had his headquarters, were also seeing the benefits of the industry and were happy to turn a blind eye to the less savoury aspects of the operation. Other famous explorers including Sir Douglas Mawson and Apsley Cherry-Garrard all tried to swing public and political opinion against the penguin hunters. The explorers of the Heroic

Age idolized penguins and brought them to the forefront of public consciousness. Public and political opposition was growing and, by the end of the First World War, pressure was mounting on the Tasmanian authorities. It was Frank Hurley, the Australian Antarctic photographer from Shackleton's expedition, that drove the final nail into the coffin. In an eyewitness account to the press, he stated, *'The birds are driven along the pens or runs and right up to the top of the digestor. Near the top of this boiler a man stands with a club, and as each bird reaches the top he hits it over the head and so knocks it into the boiler. Owing to the hardiness of most of the birds this blow only stuns them, and many go into the boiler alive.'*

Whether Hurley's account was true or false, the shock and disgust it provoked, of boiling alive these charismatic animals, was the death knell of Hatch's operation. The Tasmanian government revoked his licence and Hatch, now an old man, was forced to dissolve his business.

Over the thirty years of operation, the penguin digesters rendered the oil from around 3 million penguins on Macquarie Island. King penguin numbers had dropped from half a million to around 4,000, and royal penguin numbers had also plummeted. Today, thankfully, the numbers of both species have recovered and all that is left to bear witness are the digesters. They still stand there, right amongst the penguins, a rusting and haunting testament to a brutal and bloody past.

On other sub-Antarctic islands, many commercial companies took a heavy toll on the penguin colonies. In the Falkland Islands, over 400,000 birds were killed for their oil, almost driving the king penguin population to total

Penguin digesters

extinction. On Heard Island, another small rocky island south of New Zealand, the king penguin was hunted to extinction. The population has never returned.

Another penguin that was heavily impacted by human trade was the Humboldt penguin. This species of banded penguin lives on the western coast of South America, feeding on the multitudes of anchovies in the rich, cold Humboldt Current, from which it derives its name. Historically, this species had the unsavoury habit of digging its breeding burrows into the deep deposits of guano that other seabirds left behind. On the islands off the coast of Peru and Chile, the rich, productive waters sustain a wealth of birdlife, with millions of cormorants, boobies and pelicans feasting on the ocean bounty. The area adjacent to the Atacama Desert is very dry, and over time the white smelly deposits from the birds built up in towering hills of excrement up to 60 m deep. These small islands became known as the Guano Islands. Like all species of banded penguins, the Humboldt penguins are burrowers and found these deposits were easy to burrow into to make their nests. Most of their population breeds in this dusty, caustic environment, along with the other seabirds. Unfortunately, in the 1800s, it was discovered that the nitrous-rich guano made a superb fertilizer. Over the next few decades, guano mining became big business, and the trade was so profitable it led to wars and trade embargos. (A rather bizarre sub-plot is that the search for new guano islands started the USA down the path to becoming a colonial power.) The mining flattened almost all of the islands, extracting the guano down to the bare rock, reducing the elevation of the islands almost to sea level and totally removing the penguins' burrows. The eggs in the burrows were often eaten by the miners too, and the Humboldts, which are one of the most skittish of all penguins, failed to breed and moved away from the industrial disturbance. Over time, the miners removed all the guano from the islands, leaving the penguins homeless. It is thought that the population of Humboldt penguins dropped by 90%. Evicted from their islands, the penguins changed their breeding habits to breed on cliffs and in sea caves, burrowing in soil rather than in the guano that was no longer there. Eventually seeing the damage to their native birdlife, the Peruvian government acted to protect the Guano Islands from mining. There are still a few areas in Peru and Chile where small areas of guano remain and the Humboldts live, but these sites are now closely monitored and protected from human exploitation.

Post-war penguins

By the mid part of the 20th century, the direct commercial exploitation of most penguins had ceased. Egg hunting had become less acceptable, although it was only made illegal on the Falkland Islands in 1999. In Antarctica, in the post-war period, heroic private exploration had been replaced with more organized government-sponsored surveys and science. Overwintering bases and research stations were set up by many nations. By the 1950s, Britain had about a dozen permanent stations, most of which were around the Antarctic Peninsula, which is a haven for many species of penguin. As for the early explorers, fresh rations at these remote outposts were rare, and the local wildlife was often used to supplement and add variety to the routine menus of dried and tinned foods. The seals and penguins were being killed anyway to feed the sledge dogs, of which each base often had several dozen to aid travel over the ice on their surveying missions.

Penguins were not usually the favourite dish for the base occupants. Seals were considered much more palatable food for humans (one particular favourite dish was scrambled seal brain on toast!), and Antarctic shag, a type of cormorant, was also regarded as a delicacy. One overwintering cook, Gerald T. Cutland, working for the Falkland Islands Dependencies Survey (FIDS), the forerunner of the British Antarctic Survey, made a record of his recipes, which were later published as a cookbook, *Fit for a 'FID' or how to keep a fat explorer in prime condition.* He devoted a whole chapter to the cooking of penguins, although it is fair to say that he did not enjoy the thought of cooking them. He said that he imagined that they were just like small, inquisitive people in black-and-white coats and was worried that he would have nightmares about killing them. Neither did he eat them himself, although it appears that his colleagues on the base developed a taste for his gourmet recipes of penguin meat and other local delicacies, so that he ended up cooking wildlife several times a week. Cutland's main tip was how to prepare the food. Penguin meat smells quite badly, so here are some of his tips to get rid of the stench.

1. Wash thoroughly and leave to dry. If this does not work...

2. Hang it outside for a few days where plenty of fresh Antarctic air can get to it. If this does not work...

3. Wash it again. Then blanch by bringing to the boil in water with a little vinegar and cool immediately in cold water and wash again in cold water. If this does not work...

4. Throw it out of the window!

The recipes included: Roast penguin breast; Tornadoes of Penguin – Portuguese Style; Braised penguin breast; Casserole of penguin breast; Escallopes of penguin; Roulades of penguin breast; Savoury penguin breast; Sautéed penguin breasts; and Fried penguin breasts.

Eating penguins on Antarctic research stations was still acceptable well into the 1980s, especially as an emergency food. The Australian National Antarctic Research Expedition (ANARE) recommended penguins as food in a crisis, giving advice on how scientists should dispatch them, suggestions which included sitting on the poor beasts to squash the air from their lungs until they suffocated. Unlike Cutland's more gourmet recipes, ANARE suggested a simple penguin stew for the hungry research staff.

By the end of the century, public opinion and the growing number of research stations meant that the slaughter of seals and penguins for dog meat was no longer acceptable. In 1991, the Antarctic Treaty partners signed the Protocol on Environmental Protection, banning dogs and other nonnative animals from the continent. Needless to say, killing penguins and seals for human food was also frowned upon and today, as far as we know, no one kills penguins for food, or for any other reason.

Penguin science

Compared to the relatively brief time that we have known about penguins, the science conducted on the beasts has quite a distinguished history. Early penguin science, like most research in the 18th and 19th centuries, consisted mainly of taxonomy and descriptions of the species. This generally involved shooting the birds, bringing them home to be stuffed, then placing them into glass cabinets in museums. You can still go to many museums and find these sad and dishevelled exhibits.

The penguins of Antarctica, especially the emperor penguins, seem to have fascinated many scientists. The emperor endures some of the harshest and most extreme conditions on earth, and understanding how a warm-blooded animal can tolerate and thrive in such a severe environment has long driven hardy researchers to brave those frozen wastes to study it. The first, and possibly most famous, account of penguin science must be Apsley Cherry-Garrard's memoir of his ill-fated journey to recover an emperor penguin egg. Around the turn of the 20th century there was a hypothesis, called the theory of recapitulation, that an embryo inside an egg followed the evolutionary history of the species. It was thought that the emperor penguin was the most primitive of birds and that the embryo inside its egg would be so primitive that it would resemble a tiny dinosaur, or at least have the teeth of a reptile. Unfortunately, no one had ever managed to collect an emperor penguin egg, as the birds laid them in the darkness and extreme cold of the Antarctic winter, so getting one before it developed was not an easy task. But when Scott overwintered on Ross Island in Antarctica in 1911, just before his fateful and tragic race to the Pole, three of his party, Cherry-Garrard, Edward Wilson and Birdie Bowers, hatched a plan to visit the colony on the other side of the island to collect an egg and test out the theory. The breeding colony had been discovered on Scott's previous expedition and was less than 50 miles, as the crow flies, from Scott's hut. The book, *The Worst Journey in the World*, famously catalogues the trials and tribulations of the three men and their near-disastrous trip to collect the precious eggs. The men were ultimately successful in getting the eggs, but on returning the fragile specimens back to Britain, it was found that the embryos did not look like miniature dinosaurs, and their heroic and almost catastrophic effort did little to advance the science of embryology.

There has been a great deal of science done on emperor penguins since. Indeed, their diving behaviour may be some of the best understood of any wild bird. One of the pioneers of penguin science, Gerry Kooyman, was the first to fit automatic dive recorders on to penguins to study their diving behaviour, in 1971. The original crude devices measured just the maximum depth of the dive, and had to be retrieved by hand, but these instruments were the first ever to record the dive depth of any seabird. Improvements followed, and soon smaller, more advanced devices existed, to study the dive duration, speed, movement and other information.

One of Kooyman's protégés was a pioneering cardiologist called Paul Ponganis, who juggled two jobs, one as a heart surgeon in California and one as a penguin scientist. Throughout the 1980s and 1990s, Paul, and a team from Scripps Research Institute, travelled down to Antarctica to conduct groundbreaking research on the emperors. Using a helicopter, each year they would capture ten adult, non-breeding emperors from their colony and move them to a wired enclosure on a remote part of the sea ice, where a team of researchers waited to conduct experiments on them. After being chemically anaesthetized, the animals were fitted with time-depth recorders and backpacks that held video cameras and were wired up with heart rate monitors and blood oxygen equipment. They were then let loose to feed in the Southern Ocean. The enclosure where they were kept, known by the scientists as the 'Penguin Ranch', was cunningly located on thick sea ice, with only one hole in the ice, so the penguins had to come back to camp to breathe and could not escape from their cage. At the end of the summer, all the birds were taken back to the colony and released unharmed. One can't but wonder about the tales of alien abduction that these poor birds could recount to their fellow penguins!

The heart rate monitors and dive information revolutionized our understanding of how these birds collect their food and illustrated the incredible adaptations of the species. The original dive loggers were adapted to be used on many species, at many sites. But one drawback was that the more advanced the loggers got, the more expensive they became, and each one had to be retrieved by finding the penguin once it came back to land. As all the penguins look the same, and there are thousands of individuals in a breeding colony, finding the right penguin with your instrument on it can be problematic. Many species spend days at sea, so imagine waiting for days and days in the cold, harsh conditions of Antarctica, hoping that your penguin will come back and that it hasn't been eaten by a leopard seal! The answer to this problem came with the development of radio transmitters to help find the penguin and retrieve the precious data. These remote devices also enabled scientists to track the birds and map their foraging trips. This radio telemetry was first applied to chinstrap and gentoo penguins by husband-and-wife team Wayne and Sue Trivelpiece, on the Antarctic Peninsula. It was a technology that was subsequently used for many other species of penguin and seabird. Today, tracking devices are regularly used on almost all species of penguin to find out where they forage. Modern devices are small, and have GPS transmitters that can ping data up to a satellite and record information on penguin behaviour wherever it goes, anywhere on earth.

Other more recent advancements are DNA analysis

and gene sequencing, which are used to investigate what penguins eat, and their ancestry, while ingenious inventions such as mechanical weighbridges have been developed to automatically collect data on weight and foraging trips. Scientific information on foraging locations, diet, rates of mortality and breeding success, as well as interactions with humans, are all vital pieces of information to help conservationists protect these animals. Much of the information in this book comes from such ground breaking science, due to scores of individuals and teams from all over the Southern Ocean who have braved some of the harshest conditions on earth to expand our knowledge about penguins.

Of course, it is much easier to study penguins that live in more temperate climes, and closer to people, than those that live in the harsh conditions of Antarctica. Lancelot Richdale was originally a schoolteacher but in his spare time he studied the seabirds around his home in Dunedin, New Zealand. He conducted one of the most detailed-ever penguin studies, on the yellow-eyed penguins on the nearby Otago Peninsula on South Island. For eighteen years he watched the behaviour of the birds. He put bands on hundreds of individuals to enable him to tell each individual apart and spent long periods watching and taking notes on their courtship displays, squabbles, pair-bonding and breeding. His book, published in the 1950s, *Sexual Behaviour in Penguins*, noted down the complex 'wooing', 'marriages' and even occasional 'divorces' when a pair-bond didn't work out. It was one of the first long-term studies on penguins and became a seminal work in behavioural ecology. When I started as a penguin scientist twenty years ago, putting bands on penguins was a hot topic. For decades, scientists had used metal flipper bands to identify penguins, but a number of studies over several years had shown that the bands were detrimental to the bird's health. By comparing the penguins that had flipper bands to those without, it was found that those with the bands died younger and didn't breed as successfully. Other studies reported that it depended on the type of band, or the species of penguin. There was a great deal of disagreement in the penguin community and the debate became heated and very polarized, with scientists and institutions at loggerheads with each other. Today very few, if any, scientists use flipper bands and penguin identification is done by attaching a small transponder to their leg.

One piece of data that conservationists and wildlife managers regularly require is population counts, to see how many birds are at each site and to ascertain whether the population is growing or declining (the trend). Getting this information is fairly easy for some colonies where the birds breed close to human habitation in small numbers out in the open. The usual way to count is to use two manual observers with clickers, counting every penguin and comparing results. But the process can be challenging for banded penguins, which make burrows, or for species that breed in difficult places, such as the Fiordland penguin, which lives in the most remote forested part of New Zealand, or the Snares penguin, which only breeds on a small archipelago of remote and almost impossible-to-access rocks in the Southern Ocean. Another problem is the sheer size of penguin colonies. Many colonies of brush-tailed and crested penguins are thousands strong. Sometimes there are over 100,000, or up to 1 million, breeding adults at a single site, which makes a manual count impossible. In the early days, scientists would just 'eyeball' a colony and make a guess at the numbers. This advanced to walking around the edge of the nesting birds and working out the area, then multiplying by how dense the colony was. Today, high-tech solutions like drones and artificial intelligence are often employed to estimate the number of birds. Another option is a technique called mark-recapture. For penguins, this involves marking several birds before they go out foraging, then counting the birds coming back, to see how many birds you count before you see the ones that you marked. If you do this enough times it gives you a good idea of the total number of birds in the population.

While these methods are all right for colonies that you can access, some Antarctic species, like the Adélie and emperor penguins, live in such remote regions that a visit by scientists on the ground is pretty much impossible. Many sites have never been seen, even from the air. In these cases, high-resolution satellites have been used to find the colonies and count the birds. Satellites, even the best ones, don't have the resolution of drones and generally give coarser population estimates than ground or aerial counts, but are fairly cheap in comparison and extremely useful for monitoring long-term population trends and assessing distribution, especially if you cannot get to the location any other way. Over half of the sixty-six known emperor penguin colonies have been found using satellites, and many of these have never been seen by the human eye, even from a plane, only from the sensors on satellites orbiting the earth. My personal expertise is mainly on finding and counting penguin populations by satellite. It means that I do not get to visit penguins as often as some of my scientific colleagues, but it does have one advantage: you cannot smell them from space!

Penguin tourism

For many people, their introduction to penguins, and often their only face-to-face contact with them, is in a zoo. While this is an excellent way to view the animals close up, it cannot compete with seeing the birds in their natural environment. For those of us who love penguins, seeing them in the wild is a totally different experience. As there are many different species of penguin in many different countries, the ease of getting to a penguin colony, and the experience you get when you get there, can be very different. The penguins that live in more northerly and populated regions are usually easier to access. Two species, the little blue penguin in Australia and New Zealand, and the African penguin in South Africa, are particularly urban and often cohabit very close to human populations. If you are ever in these locations, seeing them is relatively easy and a well-developed tourist industry has established at many locations in these three countries.

Little blues spend all day in the ocean and come ashore at sunset or in darkness. They do not seem to mind being watched by crowds of tourists as they waddle up the beach on their way back to their burrows. At some locations, grandstands have been built with floodlights to better show the spectacle and to prevent over-inquisitive tourists from interfering with the bird's daily commute. Over time, it seems that the penguins get used to the human voyeurs, and it makes little or no impact on breeding success or populations. Several studies on the impact of tourists on penguins show that as long as people stay about 5 m away from them and make no sudden movements, most birds will just carry on, oblivious. Some species are a little more cautious and sometimes penguins that have not become habituated to interactions with humans will be skittish, and disturbance can occur, which can be bad for the chicks, especially if there are hungry predators around.

And penguins are a big draw. Phillip Island, a popular destination for penguin fans near Melbourne, on the south coast of Australia, hosts around 3.5 million tourists per year. In South Africa, Boulders Beach hosts 60,000 visitors each year, to see the African penguins breeding there. The Galapagos Islands have well-known penguin tours to view the rare penguins there, and some tours offer diving and swimming experiences. But if you want to see larger numbers, join the other 100,000 visitors that head each year to Punta Tombo in Argentina to see the world's largest Magellanic penguin colony.

Further south, the Falkland Islands in the South Atlantic are an excellent location to see penguins. Here, five species (king, gentoo, rockhopper, Magellanic and macaroni) breed, and this is probably the largest diversity of penguins in one place. Getting around the Falklands however is not easy, with many tourist sites only accessible by a small plane journey out to the various islands that host some of the greatest congregations of seabirds anywhere on earth.

Further east, South Georgia hosts some of the greatest

density of birds, with almost half a million kings, and huge populations of gentoo, macaroni and chinstrap, mixing on the beaches with vast numbers of fur and elephant seals. It is said that the coastal strip of this small island has the highest biomass per square kilometre of anywhere on the planet. Quite a spectacle.

For most people, when we think of penguins, we think about Antarctica, and seeing penguins in the remote pristine beauty of the white continent is often a highlight of any wildlife spotter's list. A growing number of tour ships visit Antarctica each austral summer and most of the punters on board are there to see the penguins, seals, whales and jaw-dropping scenery. Tourism in the region started on a small scale only a few decades ago. The first tourist boat to travel to the continent was in 1966 and a few years later, in 1969, the first cruise ship offering sightseeing tours crossed the Drake Passage on its way to visit the Antarctic Peninsula. Today, the Antarctic cruise industry caters for around 200,000 guests each season. Almost all of them head out from South America across the Southern Ocean to the Antarctic Peninsula, where they aim for a few historic sites and a multitude of Adélie, chinstrap and gentoo penguin colonies. The penguin colonies here can be massive, often several hundred thousand birds at a single site. On some islands it seems like penguins take up every square metre of land. Antarctic tourism is heavily regulated through the Antarctic Treaty system, providing broad rules and guidelines for tourist activity. Much of the tourism industry itself falls under the International Association of Antarctica Tour Operators (IAATO), who advocate for safe and responsible travel to the Continent. IAATO members follow rules at Antarctic tourist sites, which ensure that over-eager tourists do not disturb the birds and not too many guests visit sites at any one time. If you do want to visit Antarctic penguins it is better to choose a small ship, as landings are often limited to 200 people, and you will get a much better experience and more time on land from a smaller boat.

The ultimate experience for many tourists, whether first-time explorer or seasoned birder, is to visit an emperor penguin colony. This is not an easy feat. Only a few specialized ships, with ice-breaking capability, can get through the fast ice to Snow Hill Island, the most northerly and just about the only emperor breeding site that ships can reach. Even then, getting to the colony is not guaranteed; a change in the wind conditions can shift the pack and force even the strongest ships to run for cover. In 2009, the *Kapitan Khlebnikov*, a Russian icebreaker converted into a tourist ship, and at the time the strongest and most capable ship in the Southern Ocean, was trapped in the pack ice for over a week, on its way to the colony. But, if you do take a chance and get there, it is definitely worth it. The documentaries don't lie, and a visit to see his imperial majesty the emperor will stick in your memory for a lifetime.

But travelling to Antarctica to see penguins can be harmful. Maybe not directly: if well managed, the disturbance and other potential physical impact can be negligible, but you should also think about the indirect cost. Going to Antarctica by ship or plane entails a huge carbon footprint. Luxury icebreakers and large cruise liners burn massive amounts of fossil fuels, and we know that climate change caused by rising carbon emissions is already having a significant effect on several species of penguin. In this roundabout way, Antarctic tourism is contributing to the demise of many of the penguins that it bases its popularity on. It would be hypocritical of me to say that people should not go to see them, but if you do, think about how you can reduce your own carbon footprint. Maybe try to offset some of the carbon from the trip or do something that will make a difference back at home. Perhaps you could use the trip and your experience to become an advocate for these amazing birds.

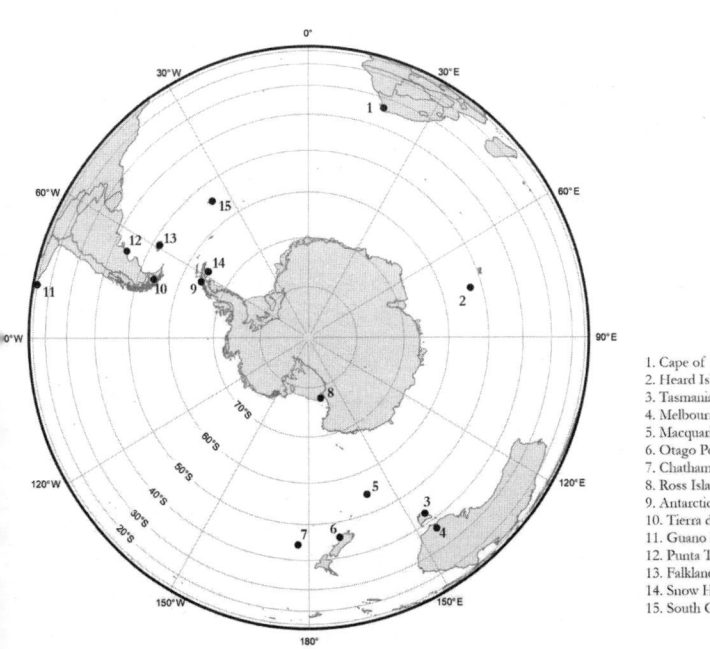

1. Cape of Good Hope
2. Heard Island
3. Tasmania
4. Melbourne
5. Macquarie Island
6. Otago Peninsula
7. Chatham Islands
8. Ross Island
9. Antarctic Peninsula
10. Tierra del Fuego
11. Guano Islands
12. Punta Tombo
13. Falkland Islands
14. Snow Hill Island
15. South Georgia

Penguins in culture

Penguins have charisma, people like penguins and penguins sell, so, over time, these cute little birds have infiltrated human culture like very few other animals. But this isn't a new phenomenon; ever since penguins were discovered, they have become something of a favourite. The first penguins on view to the public would have been stuffed animals in museums, often sad-looking affairs, but unusual and rather comic nonetheless. These specimens were soon followed by real live birds in zoos. Temperate penguins are relatively easy to keep in zoos, or at least in zoos with a relatively cool climate, and today are one of the mainstays of any major zoological attraction. Edinburgh Zoo is credited with having the first live penguins when, in 1913, a whaling ship, travelling back from the Southern Ocean on its way to Norway, docked at Edinburgh and gave three live king penguins to the newly built zoo. The birds were thought to be the first penguins on public view anywhere in the world. Edinburgh Zoo still has an association with penguins and has kept a number of king penguins as an exhibit there ever since. In a rather bizarre substory, one of their kings has become famous as the mascot of the King of Norway's Royal Guard, and this particular military penguin, named Sir Nils Olav III, Baron of Bouvet Island (a small rocky island in the Southern Ocean claimed by Norway), has been knighted and given the rank of Major General by the Norwegian Army.

Several major zoos followed Edinburgh's lead and brought penguins north. The timing of these first zoo-based birds early in the 20th century corresponded with the Heroic Age of Antarctic exploration, when the exploits of Amundsen, Scott, Shackleton and many other polar explorers became the talk of the age. Their stories were made even more popular through touring exhibitions of photographers such as Herbert Ponting and Frank Hurley, where images of penguins often featured prominently in their shows. Penguins were a perfect photographic subject and, even with the slow and cumbersome camera equipment of the day, the curious little birds were great at posing for the lens. Even if they didn't, the cameramen often had tricks to get the perfect shot. A famous penguin photo from the Sottish National Antarctic Expedition in 1904 shows a picture of a Scotsman playing the bagpipes, in a kilt, in front of a bemused-looking emperor penguin. But if you look closely, you can just see the rope that is tied to the foot of the piper and attached to the leg of the penguin, to stop it running away. The photos from these Antarctic expeditions did a great deal to raise awareness of penguins and start to cement them in the public psyche.

Not all zoos found it so easy to adopt penguins. King penguins prefer cooler temperatures, which is fine for chilly Edinburgh but not so good for warmer cities, so penguins from more temperate climes, such as African penguins or Humboldts from South America, were often preferred. London Zoo built its first penguin pool in 1934, and no expense was spared for the plush penguin pad. The Bauhaus architectural marvel of a pool won several design awards and the zoo was successful in keeping a number of Humboldts to delight the crowds of London. The first penguins in America, at the Smithsonian National Zoo in

Washington, proved difficult to keep alive and required high maintenance, probably due to the hot, humid summers on the east coast of America. After several failed experiments at keeping various species in the 1930s, a group of African penguins were eventually successful in raising two chicks and, although both died before fledging, tens of thousands of people showed up to see the new arrivals.

Today, although there are hundreds and hundreds of zoos that keep penguins, the cost and high maintenance of the birds is offset by their huge appeal. But very few zoos keep Antarctic penguins, which prefer constant cold temperatures, and only one or two keep emperor penguins, as they are very difficult to acquire, and even harder to keep. Although many penguins breed in captivity, emperors are extremely hard to breed. San Diego Zoo recently reported that they had been successful in raising a single chick from their seven adult emperors after thirteen years of trying.

By the middle of the 20th century penguins started to get into literature too. Possibly the first famous book about them, written in 1938, was *Mr Popper's Penguins*, a story where an ordinary household inherits first one and then a whole family of penguins.

However, this fictional tale had been preceded four years earlier by possibly the most famous penguin in literature: the penguin logo of Penguin Books (see page 246 for a brief history of Penguin Books). Created by the publisher Allen Lane in 1934, the brand introduced affordable, quality paperbacks, synonymous with the symbol of the penguin. The symbol is actually a Humboldt penguin drawn from one of the animals recently installed in London Zoo. For ninety years the jaunty black-and-white symbol has been on the spine of millions of books sold worldwide. The tale is told that it was Lane's secretary who suggested that the penguin would make a good logo as the birds were *'dignified but flippant'*.

By the 1940s the first cartoon penguins appeared on our screens in a segment of the Disney film *The Three Caballeros*. Part of the film tells the story of Pablo, a penguin who finds Antarctica too cold and seeks the warmer weather of the Galapagos Islands. Pablo is thought to have been the inspiration for the first famous early cartoon penguin, Chilly Willy, a small, cheeky and likeable character who starred in fifty short films for the Walter Lanz studio between 1953 and 1970.

The initial detailed documentaries of penguins included the David Attenborough-voiced BBC film, *Life in the Freezer* from 1993, which showed the seasonal cycles of the frozen continent through its wildlife. But it was not until the start of the 21st century that penguins in films really started to proliferate. One of the really big hits was the 2005 feature-length documentary *March of the Penguins*, which showed the extraordinary breeding cycle of the emperor penguin. The production, made by a French team at the research station of Dumont D'Urville in east Antarctica, highlighted the trials and endurance of this remarkable animal. At this research station, the French scientists had been studying the neighbouring emperor colony for decades, and it was one of the very few places from where you could access the breeding site in the freezing dark of the Antarctic winter. Watching the film for the first time, the brilliant camerawork makes it look like the emperors are in the icy wilderness far from human habitation, but I have spoken to French scientists who say that they could sit on the veranda of their scientific station, sipping their coffee, and watch the penguins, sometimes just 100 m away. The film, voiced in English by the actor Morgan Freeman, won an Oscar for best documentary, had critical success and became one of the top-earning documentary films ever made.

And then there was Mumble. I cannot say that I am a fan of the tone-deaf dancing penguin in *Happy Feet*, but he certainly made an impact, with the film being one of the highest-grossing films of 2006 and winner of the Oscar for best animated film. The comic cartoon emperor chick spawned a sequel, and a host of other animated penguins soon followed. In his wake, who can forget the savvy penguins from the film *Madagascar*, Skipper, Kowalski, Rico and Private, who were so popular that they got their own feature-length film. Or Cody Maverick, the surfing rockhopper from *Surf's Up* or, a personal favourite of mine, the dastardly villain Feathers McGraw in the Wallace and Gromit hit, *The Wrong Trousers*, and, more recently, his dastardly revenge in *Vengeance Most Fowl*. Hollywood's love affair with penguins has led to at least seven full-length feature films starring these delightful little critters over the last thirty years, making them the most popular wild animal on the silver screen.

The small screen has also seen its share of black-and-white flippered stars, mainly in the form of the tooting plasticine Pingu, who always seems to get into trouble. There are many other cute and fluffy TV penguins on offer to children around the world, although of course these cute, feathered characters were all preceded by The Penguin, the dinner-jacket-wearing, top-hatted, despicable arch-enemy of Batman in the original DC comics and *Batman* TV series of the 1960s, more recently returned in his own dark crime drama series for the streaming service HBO.

Over time, penguin popularity has become big business. In fact, Penguin Books' successful logo was not even the first recorded marketing use. Penguin biscuits predate Penguin Books by a couple of years. The chocolate-covered treats were launched in Scotland in 1932 by Macdonald Biscuits. They became a much more household name when McVitie's bought them out in 1946. Mass marketing made them a British icon with the famous line, '... *if you are p-p-p-peckish, p-p-p-pick up a penguin*'. Today, McVitie's UK factory makes between 200 and 500 million penguin bars each year. Other notable uses of penguin logos include the Pittsburgh Penguins ice hockey team, and Tux, the icon of the Linux computer operating system. Another example of the power of penguins is Munsingwear, a Minnesota-based clothing company. Originally specializing in underwear, their wool-silk mix long johns were a particular favourite for chilly winters in the Midwest. They have been making clothing since the 1880s, but it wasn't until the 1960s, when they branched out into sport and leisurewear and put a tie-wearing penguin symbol on their shirts, that the brand made it big. Today the 'Original Penguin' brand of clothing is a multimillion-dollar brand.

In advertising, penguins are commonplace. Their trait of liking the cold means that they are perfect for marketing anything frozen, while their cute, comic nature makes them endearing fluffy friends or novelty characters. Film, TV, advertising, literature: for an animal that lives so far away, and that many people will never see in the wild, this black-and-white superstar has become ubiquitous in global culture.

How to say 'penguin' in sign language.

Chapter 6:
SPECIES

Chinstrap penguin
Pygoscelis antarcticus

Height: 68–76 cm, **Weight:** 3–5 kg
Range: Sub-Antarctic islands and northwest Antarctic Peninsula
Population: 8 million, **Conservation Status:** Least Concern
Diet: Almost exclusively krill

The chinstrap penguin is named after the smart line of black plumage that runs under its chin and which, along with its black cap, looks like the strap of a helmet. It has a loud, harsh call that has led it to be named the 'stone-breaker', the logic being that the sound is so awful that it could break stones. It is one of the three brush-tailed, or *Pygoscelis*, penguins, along with the Adélie and gentoo. As well as its panchromatic plumage it also has a black beak, which probably makes this medium-sized penguin the most black and white of any penguin species. The only colour it has on its body are its greyish-pink feet (less colourful than

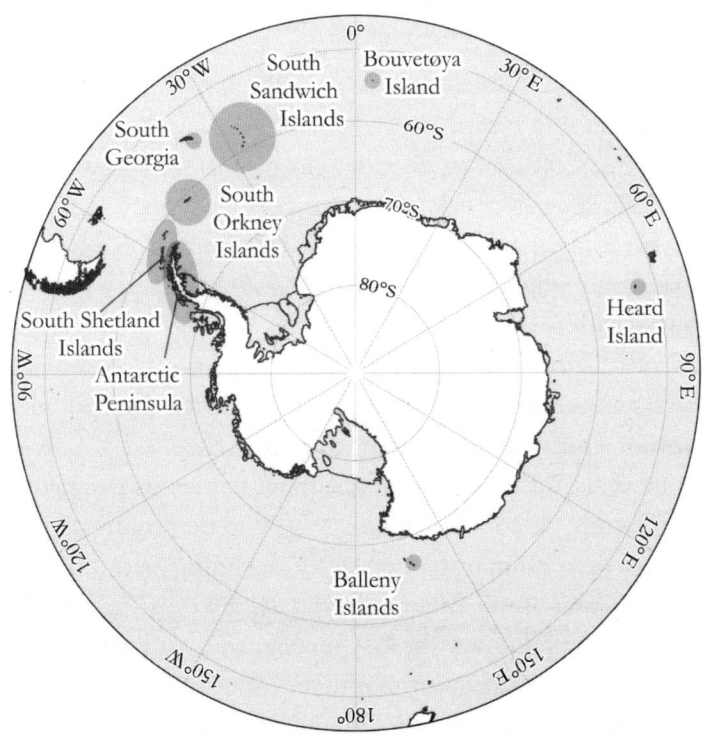

the Adélies' yellow boots) and brownish eyes. Its breeding range is fairly wide, concentrating on the islands around the Scotia Sea, with the main centres on the sub-Antarctic archipelagos of the South Sandwich, the South Orkney and the South Shetland Islands. The colonies it inhabits are often extremely large, sometimes numbering in the hundreds of thousands. The breeding site on Zavodovski Island, the most northerly of the South Sandwich Islands chain, is reputed to be the world's largest penguin colony, with up to 1.5 million birds spread over most of the small 4 km² island.

Smaller, but still substantial, populations exist on the northwestern part of the Antarctic Peninsula and there are reports of very small colonies on other, more widespread sub-Antarctic islands, such as Bouvet Island and South Georgia. Colonies often exist close to or amongst other penguin species (a behaviour that ecologists term 'sympatric breeding'); at Zavodovski, the chinstraps cohabit with macaroni penguins, while further south colonies are often mixed with Adélies. At some locations on the South Shetland Islands, all three of the brush-tailed penguins can sometimes be seen breeding at the same site. Overall, the total population of chinstraps is thought to number around 8 million individuals, making it one of the more numerous penguins.

Current scientific thinking suggests that many of the sites where chinstraps breed are declining with climate change. The South Atlantic region, where most reside, has warmed rapidly over the last few decades and there have been several records of population declines. But not all sites have the same trend. Colonies in the more northerly parts of their range, such as the South Orkney Islands and South Shetlands, seem to be declining, but more southerly colonies like those around the central part of the Antarctic Peninsula are stable, and some smaller sites are increasing in numbers. This worrying trend has been attributed to climate change-induced reductions in the supply of their main food, Antarctic krill. However, around half of all 'chinnies', as scientists like to refer to them, breed on the South Sandwich Islands, one of the most remote groups of islands on earth. Survey results from these large, isolated populations are few and far between, and trying to ascertain the population trend, when the original data is so patchy, is not an easy task.

Chinstrap colonies are densely packed and noisy. The birds are noted as being one of the more aggressive species of penguin, at least with each other, and they love a good squabble. They build large, rocky nests and raise one or two chicks, often a few weeks later in the season than the other brush-tailed species. They usually arrive at their colonies in early November, after spending the winter north of the Antarctic pack-ice zone. Chicks hatch in January, after about 35 days of egg incubation, and adults and chicks will have left the site before the onset of the Antarctic winter in April. Chinstraps tend to breed on the slopes of small hills, mainly due to the fact that they arrive at their breeding locations after the Adélies, who bag the prime real estate on the tops of the hills first. When they live on their own, they are more than happy to sit on the top of the hills, as these sites tend to become snow-free earlier in the season and are less prone to flooding. A chinny typically lives for fifteen to twenty years and will start to breed between the ages of three and seven years. They have elaborate breeding displays and usually mate for life, unless the partnership doesn't work out, in which case, like us, they will 'divorce' and pick another partner who, hopefully, will be more compatible.

They feed on krill, usually catching them fairly near the surface and, when prey is abundant, tend to forage within 40 km of their colony. But occasionally, in years when their main food is in short supply, foraging trips may be 100 km or more. This dependence on one prey species, whose distribution is starting to change with the warming of the oceans, is a concern for ecologists who think that chinstraps will be hit hard by future climate change.

Adélie penguin
Pygoscelis adeliae

Height: 70-73 cm, **Weight:** 4-6 kg
Range: Pan-Antarctic: anywhere around the coast of Antarctica where there is a rocky shore
Population: Around 10 million, **Conservation Status:** Least Concern
Diet: Mainly krill, some fish, squid, jellyfish and crustaceans

Feisty. That is the description that scientists often repeat when I chat to them about Adélie penguins. These charismatic little birds are renowned for their cheeky, no-nonsense character. Standing at around 70 cm tall, approximately the same size as a chinstrap, what the Adélie lacks in size it makes up for in personality. Although their plumage is purely black and white, they seem to have very expressive

faces, possibly due to their ability to ruffle the feathers on top of their head and widen their eyes to expose their white eye-rims when they are annoyed. They also seem to have slightly outsized flippers, which they raise high when they walk, giving them an amusing waddle. Like all brush-tailed penguins, they tend to nest in very large, dense colonies that are noisy, smelly metropolises of life.

They were named by the French explorer Jules Dumont D'Urville, after his wife Adélie, in 1840, as he made his way around the coast of east Antarctica. Indeed, it is the only penguin other than the emperor that lives on the frozen continental coasts of east and west Antarctica, south of the Antarctic Peninsula. It breeds on the Peninsula too, but here it has to share the krill, its favourite food, with its cousin brush tails, the gentoo and chinstrap penguins. In these

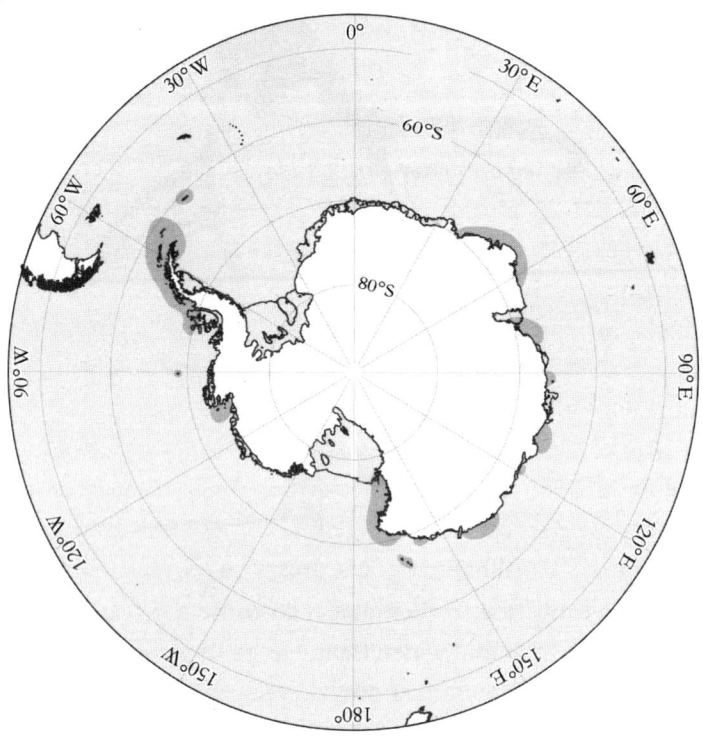

more northerly locations they often breed in close proximity to the other penguins, mixing freely with their neighbours. Adélies are better adapted to life in the pack ice than other brush tails, and, unlike them, tend to stay around the coast of the frozen continent all year round. That allows them to get to their breeding colonies before their fellow citizens and bag the best nest sites on the higher ground, away from the slopes and gullies that are sometimes prone to flooding in rainstorms or snowmelt. Returning to their breeding sites in September, they will tend to their stony nests and find their mate, then start their breeding cycle in October before the other species arrive, laying a clutch of two eggs. Their incubation period lasts around 33 days, meaning that their chicks hatch in early December. Within a month, the fluffy chicks will be big enough to leave the nest and form crèches, before fledging into their waterproof plumage at the age of 7–9 weeks old. Adults are mainly monogamous, frequently remaining with the same partner for many years. After the chick-rearing period, the adults will also leave the breeding sites, travelling deep into the remaining pack ice to do their annual moult, their once-a-year change of clothing.

As we have mentioned, most Adélies live south, beyond the Peninsula, around the bleak, frozen shoreline of continental Antarctica. Here, where the vast ice sheets meet the sea, rocky coastal habitat is at a premium and sometimes it feels like every isolated rocky patch or bare offshore island is home to a penguin settlement. These rare rock outcrops are also used by scientists as locations for their research stations, so many Antarctic bases will have an Adélie penguin colony close by, which may be the reason why this bird is one of the best-studied penguins in the world. The size of these colonies can be truly massive. The largest aggregation is thought to be on the Danger Islands where around 1.5 million breeding birds are spread out across the archipelago of seven or more small islands. The Ross Sea is a hotspot for this species, with many large colonies along the coast of this giant embayment. These sites include Cape Adare, on the northwest corner of the Ross Sea, which is possibly the largest single breeding site with, at the last count, around three-quarters of a million birds.

The population trend for the species is a mixed picture. In more northerly locations around the western Antarctic Peninsula, where the climate has warmed considerably over the last few decades, the population is declining rapidly. Here, when it gets too warm, snowfall turns into rain and the chicks' downy feathers get wet and they struggle to keep warm. Declining krill populations linked to the disappearance of sea ice in the region have also had an impact, meaning that there is not enough food to go around. Other brush tails, more suited to slightly warmer conditions, take the majority of the share. However, further south, especially in the huge colonies around the Ross Sea, the declines on the Peninsula have been offset by steady increases in the population. As these more southerly colonies are significantly larger than those on the western Peninsula, it is thought that overall the population is increasing. Here, since the turn of the century, sea ice has been increasing and so have the penguin numbers, although how long this will last with the oceans and sea-ice loss happening further and further south is a moot question.

Gentoo penguin
Pygoscelis papua

Height: Around 76 cm, **Weight:** 4.5–7 kg
Range: Falkland Islands, Atlantic and Indian Ocean sub-Antarctic islands, and the northwestern part of the Antarctic Peninsula
Population: around three quarters of a million
Conservation Status: Least Concern
Diet: Variety including fish, squid, krill, jellyfish and crustaceans

SPECIES

The third species of brush-tailed penguin is the gentoo. It is the third-largest penguin species, standing at around 76 cm tall, although it is much closer in size to its two, slightly smaller, cousins, the Adélie and chinstrap, than the larger king and emperor penguins.

How the gentoo got its name is a bit of a mystery. The leading theory is that 'gentoo' may be an old slang term used by Portuguese sailors to denote people of southern Asian descent, although a slightly different, but related, explanation is that it was a colonial British term (possibly derived from the original Portuguese) to differentiate between Sikhs and Muslims. The theory being that the white stripe on the gentoo's head resembles a turban. Personally, I cannot see the resemblance, but however the origin, we are stuck with it. The Latin name is much easier to explain, although still somewhat confusing; the *papua* part derives from 'brush-tail', the common name of the family, and the Latin name of the brush-tailed family is *Pygoscelis*, which means 'rump legged', not 'brush-tailed'. It did not help that the person who first described the penguin scientifically, Johann Reinhold Forster, the naturalist on Captain Cook's second expedition, and the same chap whom emperor penguins were later named after in their specific name, got totally mixed up and thought that people called them *papua* because they

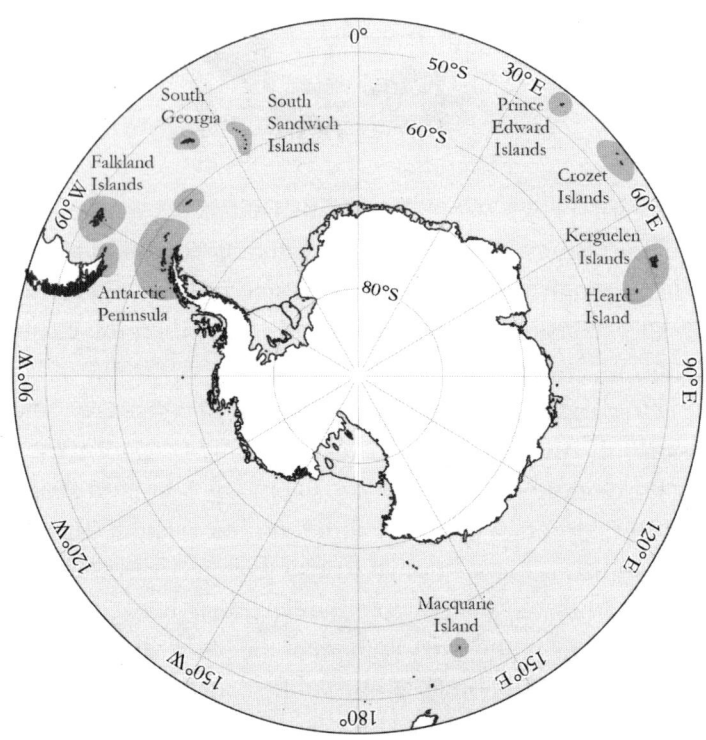

came from Papua New Guinea. Confused yet? Maybe we should move on.

Gentoo penguins are mainly black and white but can be differentiated from the other brush-tailed penguins by the rakish white flash above their eye and their bright orange beak. They are sturdy-looking birds and are thought to be the fastest-swimming penguins. They feed on a wide variety of fish, krill, other crustaceans, jellyfish and other marine organisms, a much more diverse diet than most other species. Whereas the Adélies and chinstraps will spend days swimming out to their preferred feeding ground to fill up on krill, the gentoo stays much closer inshore and fills up on whatever it can catch in the shallows, returning to shore each evening to roost at its breeding site.

Their distribution ranges from temperate regions, like the Falkland Islands, to pretty much all over the sub-Antarctic islands, down to the Antarctic Peninsula. It may be that this versatile bird has one of the greatest breeding ranges of any penguin, although those that live in warm, more northerly latitudes have been found to be considerably larger and heavier than their Antarctic brethren. When I look through my (rather large) photo collection of penguins, it also seems that there are some colour differences, with Antarctic gentoos having pinkish feet, compared to the more yellowish feet of the Falkland gentoos, and they appear to have brighter orange beaks. Scientists have started to think that what we classify as gentoos may be more than one, and even up to four, different species.

Generally, they breed in smaller groups than other Antarctic penguins, often nesting on the periphery of large chinstrap or Adélie colonies in Antarctica or further north, mixing in close to king or macaroni penguins. My own feeling is that they are a little less noisy and argumentative than the other brush tails, which live in such massive sprawling conurbations; they are positively rural in comparison to their streetwise urban cousins.

But being more versatile in their diet and not depending upon a limited number of very large breeding sites does seem to make them more flexible and it appears that, of all the penguins, the gentoo is one of the few that seems to be adapting best to climate change and the warming of the oceans. Maybe it is the fact that it is less dependent upon krill as a foodstuff. Krill numbers are thought to be linked to the sea ice, which is in rapid decline around the Antarctic Peninsula, and krill numbers and distributions may be changing accordingly. Or maybe it is just the overall versatility of the species helping it adapt. Whatever it is, gentoos seem to be the only penguin in the sub-Antarctic or Antarctic Peninsula region whose population is on the up, bucking the trend of the other penguins around it.

From a breeding standpoint, the species is flexible too. Chinstraps and Adélies are rigid in the timing of their breeding cycle, always turning up at the same time, no matter if the snow and ice melted early or late that year. But our flexible friend the gentoo will change the timing of its breeding cycle depending upon the conditions, although, generally, its chicks hatch sometime between the other two brush-tailed species, often as early as when the Adélies hatch or as late as the chinstraps, a month later. Other breeding traits are similar to other brush tails, but their shorter foraging range has its drawbacks. If food is scarce in the local area, gentoos cannot travel further to find better foraging, so occasionally, in bad years, chick mortality can be very high. They build large, stony nests that they guard jealously – did I say they were less argumentative? That is unless you steal one of their stones; then all hell can break loose, with squabbling, fighting and braying. They pair up for life, and are more faithful than almost all other penguins. It is said that a gentoo that commits adultery will be banished from the colony.

African penguin

Spheniscus demersus

Height: 70 cm, **Weight:** 2.2–5 kg
Range: Mainly South Africa, with a small population in Namibia
Population: 41,700, **Conservation Status:** Critically Endangered
Diet: Anchovies, sardines and other small fish, squid and occasionally crustaceans

The African penguin is also called the jackass penguin for its raucous braying call, or the black-footed penguin for its black footwear. It is a resident of the coasts of South Africa and Namibia, beyond the northern edge of the Southern Ocean. It breeds on small islands around rocky coasts and feeds on the masses of tasty fish that proliferate in the cold productive Benguela Current that runs around the bottom of Africa. They typically make their home close to the shore, but can settle almost a kilometre inland, where they dig burrows if the ground is soft enough or, on harder terrain, find an overhanging rock or

bush to keep the sun off the small scrapes that they build. It is one of the four banded penguins, which all look fairly similar. They are almost totally black and white, with the only other colour on their face or feet. They have a white chest, but for some banded penguins it is slightly speckled. The Africans usually have the most speckles, although this varies between individuals and is not a good way of identifying them. They have a curved black band between their white chest bib that merges into a horseshoe-shaped white area on their cheeks, with a black area surrounding their beak and extending out around their eyes. They also have a pink fleshy patch above their beak. This is the most colourful part on them and is used as a heat-exchange mechanism to keep them cool in the hot African summer. They are shallow divers and they don't tend to swim deeper than 100 m, although their specific name *demersus*, which means 'plunging', might suggest something a little more extreme. During the breeding season they won't travel too far from their colonies, which can number several thousand pairs of nesting adults.

The species is renowned for having very complex and flamboyant courtship routines, with many different moves

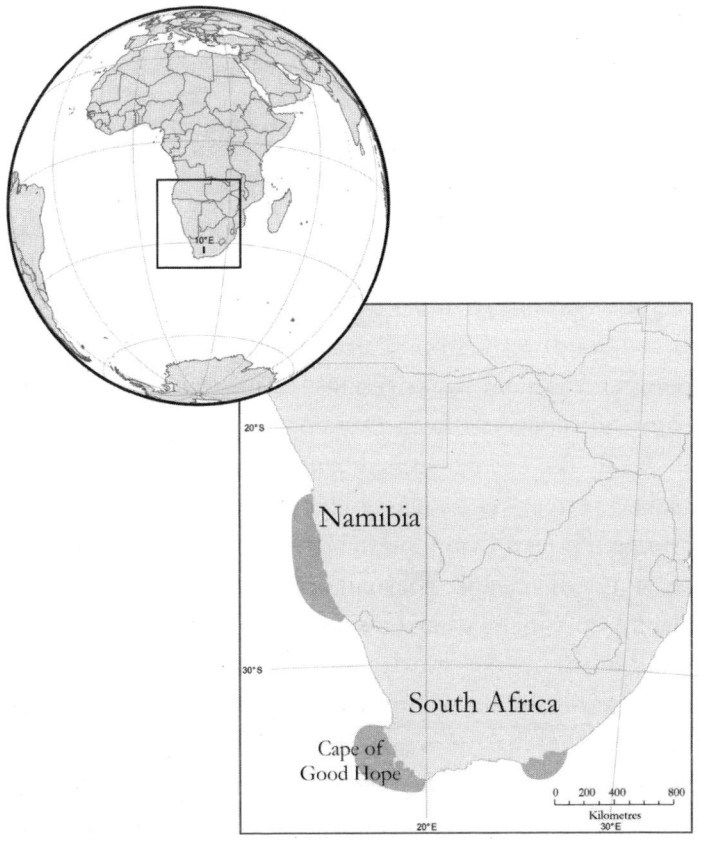

and gestures. Equally, they can be quite aggressive, and also have a variety of hostile poses to ward off other suitors.

The African penguin has suffered more than most at the hand of humankind. It is thought that before Europeans settled on the coasts of Southern Africa there were many millions of penguins living around southern Africa, but relentless hunting for meat and eggs gradually drove their numbers down. Added to this, the guano from their colonies was dug up to use as fertilizer, destroying nests and causing millions to perish. There are estimates that as many as 40 million once existed, making this possibly the most numerous penguin ever. Even after all the persecution, by the turn of the 20th century there were still about 2–4 million left, but during the past 100 years numbers have continued to drop. Penguin eggs became common food in South Africa during the 18th and 19th centuries. So common, in fact, that for many years the traditional lunch in the South African parliament was fried penguin eggs. Between 1900 and 1930 it has been estimated that 13 million eggs were collected from Dassen Island on the west coast alone! By the 1950s numbers had dropped by over 90% to around 300,000. Egg collection was banned in 1968 and the mining of guano stopped in the 1950s, mainly due to the discovery of cheaper alternatives, although in some places digging the smelly white layers for fertilizer continued until the 1990s.

However, in the last fifty years the threats have changed. Oil spills have become a major problem. The passage around the Cape of Good Hope is the busiest oil-shipping route in the world and, due to accidental spills, shipwrecks or just cleaning out tanks at sea, the death of penguins in this area has increased dramatically. There have been seventy-one major oil spills recorded in South Africa between 1970 and 2021. Thousands of birds can be affected even by a small spill and, for most, it is lethal. Oil removes the waterproofing on their feathers, making them die from the cold. If they do manage to preen it off, most expire through poisoning from the oil they have ingested. As the whole population live on this busy coast, all within 100 km of a port, they are directly in the firing line, and the latest data suggests that 2% of the entire remaining population is oiled annually. Over the last few decades active conservation networks have been set up to clean oiled birds, so that now 86% of rescued birds are washed and successfully rehabilitated back to the wild. Action to prevent spills and to rescue penguins did seem to help for a while and numbers stabilized, but about ten years ago, with the population already at breaking point, numbers once more started to decline rapidly. In 2019, there were only around 19,000 pairs left, 70% of which were in South Africa and the rest in Namibia. Scientists looked for the cause of the most recent decrease and it seems that the most likely culprit is an increase in fishing activity. The tasty sardines and anchovies on which the African penguins feed are big business, and with modern fishing methods, overfishing has depleted the penguins' food supply. In 2023, the South African government installed a temporary fishing exclusion zone around all the remaining penguin colonies, which might be their best chance of survival. If not, it is estimated that this, the most romantic of penguins, may go extinct within the next ten years.

Magellanic penguin
Spheniscus magellanicus

Height: 70 cm, **Weight:** 2.7–6.5 kg
Range: Southern South America and the Falkland Islands
Population: 2.2–3.2 million, **Conservation Status:** Least Concern
Diet: Various small fish, crustaceans and squid

The Magellanic penguin looks very similar to its African cousin. The main difference is that the white upside-down U-shaped band that surrounds their white chest bib does not merge into the white cheek patches on their face, so they have two white bands between their beak and their chest. Their feet are also somewhat more pink, but other than that they are pretty similar. Like the African penguins, the Magellanics have complex courtship routines and a donkey-like braying call, and they forage close to their colonies, chasing the small fish that are their preferred prey, quite close to the surface.

The geographic range of the Magellanic penguin is around the southern coast of South America, which means Argentina on the Atlantic side and southern Chile on the Pacific, as well as around the Falkland Islands. They are

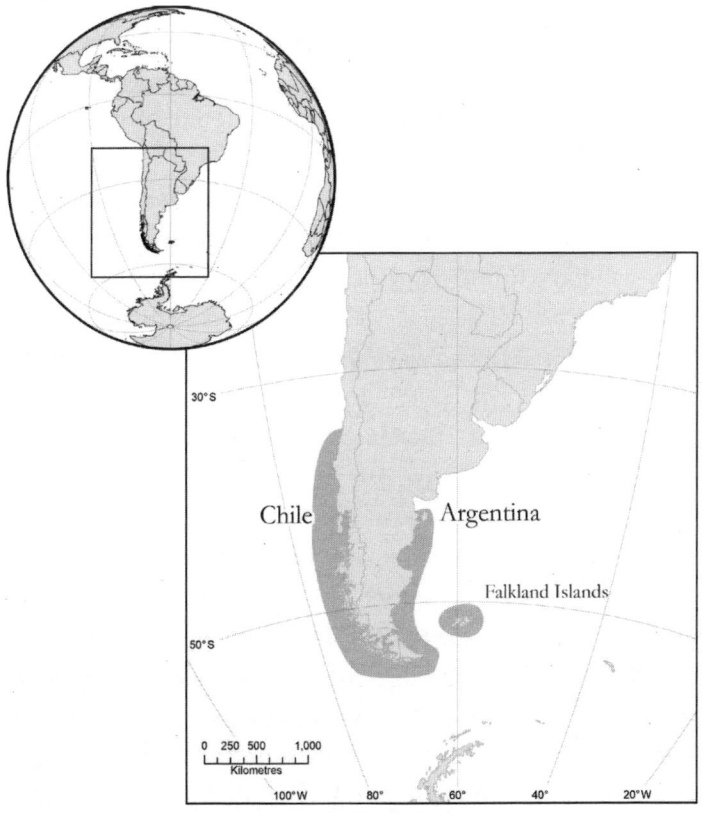

named after Ferdinand Magellan, the Portuguese explorer whose expedition first found penguins around those coasts.

They breed in large colonies on a variety of ground types. Unlike most penguins, the Magellanic does not seem to be that picky about where it makes its home. One of the classic old penguin textbooks, Tony William's *The Penguins*, quotes: *'Breeds on bare, grassy, bushy or forested islands and coasts, cliff faces and escarpments, and flatter areas, taking advantage of local vegetation where available, nesting on the surface or in burrows'*. So that's pretty much anywhere they can find! Some of the colonies are huge. At the most famous one, around Punta Tombo in Argentina, you can find around 1 million birds on one small 3-km-long peninsula, making it one of the largest of all penguin communities.

Like most penguins, their breeding cycle is tied into the short summers of these southerly latitudes. By late October they will have found a partner, and a few weeks later most females lay two white eggs, with the parents sharing the incubation duties. After 41 days the chicks hatch, then 4 weeks afterwards they will be left alone while both parents forage for food at sea. Once the chicks have become self-sufficient in April, the adults leave for their winter vacation. Many penguin species move away from their breeding colonies in the winter. For instance, the brush tails move on to or away from the sea ice when the Antarctic winter sets in, while some of the other banded and crested species travel hundreds of kilometres into the ocean to find food. But the Magellanic is probably the most migratory of all species, as each winter the birds from the Falklands and southern Argentina move long distances up the coast of South America as far as Uruguay and southern Brazil, following the shoals of anchovies that make up the bulk of their diet. In the spring, they will head south as the weather warms to get back to their breeding colonies by September or October each year.

There are still lots of Magellanic penguins around. Unlike many other species, populations seem to be holding up, with the IUCN Red List and Birdlife International classifying them as 'least concern'. Some scientists, however, worry that a lack of recent count information may be hiding the real picture. Certainly in some places populations seem to have fallen. The huge colony at Punta Tombo has declined 40% since 1987 and, on the Pacific side in Chile, many of the northern colonies have been abandoned. A mixture of climate change, destruction of vegetation and intensive fishing seems to be having an effect, but due to the scarce monitoring at most breeding sites it is impossible to calculate the overall impact. The situation is not all doom and gloom. Some of the smaller colonies that are monitored seem to be doing OK and are actually increasing in numbers, but it is always a worry when threats like industrial fishing start to alter the populations of anchovies around the South American coastlines.

Humboldt penguin
Spheniscus humboldti

Height: 65 cm, **Weight:** 4-4.5 kg
Range: Pacific coasts of Chile and Peru
Population: 23,800, **Conservation Status:** Vulnerable
Diet: Anchovies and other small fish

The Humboldt penguin gets its name from the nutrient-rich cold-water current that flows northwards up the west coast of South America. That current in turn is named after the German geographer, explorer and naturalist Alexander von Humboldt, who first noted the unique properties of this ocean flow. This penguin is sometimes known as a Peruvian penguin, as most of the population breed along the coast of Peru. Its distribution also overlaps with the Magellanic penguin in Chile and occasionally they interbreed, which has led some researchers to suggest that they are only different subspecies of the same type of penguin. Personally, I don't see it. Morphologically Humboldts are distinct from the Magellanic, being slightly smaller with chest markings more like an African penguin. The thing that tends to distinguish them best is the fleshy area around their (slightly thicker) beak, which is much more extensive in the Humboldt. This pink patch that surrounds the front of the face is used as a cooling mechanism (as the birds find it hard to lose heat out through their insulating feathers) and, as the Humboldt lives in hotter conditions, it follows that it needs this larger fleshy area to dissipate heat. And boy does it get hot! We think of penguins living in cold climates, but the Humboldt lives in the desert, and not just any desert, but the Atacama, the driest desert on earth. In such hot temperatures, where the mercury can rise above 45°C, it helps if you can keep out of the sun, so most Humboldts prefer to nest in narrow burrows, although even the burrows get hot in this heat. Insulating eggs and chicks from the sun is one of the main challenges for this species. The males make nests lined with feathers, which may also help to protect them.

The breeding success of the Humboldt penguin is deeply linked to the current from which it derives its name. When the cold current is strong, it brings nutrient-rich waters with an abundance of fish. Additionally, it revives the coast with refreshing cool breezes and sea mists. But when the current is weak, the fish go elsewhere, the temperature rises and the penguins starve. At these times it gets even hotter and many chicks and adults can die of heat exhaustion. The fortunes of the Humboldt Current are driven by the weather phenomenon known as the El Niño Southern Oscillation (ENSO). The El Niño, or 'the little boy' in Spanish, brings the warmer waters, but when

La Niña, the 'little girl', arrives, the trend reverses and cooler waters dominate. These trends tend to happen for several years at a time, so the species can endure multiple years of poor breeding, and in those years it sometimes doesn't even try to raise a brood. When the cooler waters return, bringing the fish, the penguins get busy and will often raise several broods in a year, usually two but sometimes up to three. Unlike other species that, apart from the emperor and the nearby Galapagos penguin, tend to breed in the spring and summer, the Humboldt will raise its chicks at any time of year on these tropical coastlines, but they usually prefer the winter, when the temperature is less extreme.

In 1982/83, a particularly strong El Niño event saw massive mortality and abandonment of colonies in Chile and Peru. It was so serious that populations at many colonies fell to dangerously low levels. But in the La Niña year following that event the population bounced back, as a couple of Humboldt penguins can rear as many as six chicks in a single year; this really is a boom-and-bust species.

Before modern civilization reached this part of the world, most of the penguins nested on offshore islands, safe from land-based predators. There, over tens of thousands of years, the gulls, cormorants, boobies and penguins deposited guano which, in the dry environment, never washed away. The penguins loved to burrow into these towering deposits to make their nests. But in the 1800s people discovered that guano made an excellent fertilizer and started to mine the islands, destroying the penguins' homes (see page 107). As in the African penguin story, the numbers of these birds plummeted. Forced off their islands, many had to relocate to the mainland, but here the ground was not so forgiving, so many were forced to exchange their des-res burrows and slum it in caves or under cliff overhangs to keep out of the baking sun. Governments have now protected many of their breeding grounds and tried to fence off the ones on the mainland to stop land predators from attacking birds and eggs. For a while, populations did seem to be recovering. But with climate change, the El Niño–La Niña oscillation pattern has changed, with longer and stronger extreme oscillations. Warmer air and ocean temperatures seem to be driving numbers down. That, along with heavy commercial fishing of anchovies around the coast, means that populations, once counted in their millions, are now little over 20,000 and falling rapidly.

Galapagos penguin
Spheniscus mendiculus

Height: 53 cm, **Weight:** 2–2.5 kg
Range: Western Galapagos Islands
Population: Unknown, but estimated around 1,200 individuals
Conservation Status: Endangered
Diet: Anchovies and other small fish

SPECIES

The Galapagos penguin is the only penguin to venture north of the equator, as the small archipelago of islands where it breeds and from which it derives its name straddles both the northern and southern hemispheres. Even then, only a tiny proportion of them breed north of the equator and the most northerly nests are less than 20 km above the line. Its specific name *mendiculus* means 'little beggar', not a particularly complimentary reference. A couple of explanations have been suggested for this strange nickname. Firstly, it is possible that when sailors initially encountered the Galapagos penguin, they disliked its demeanour, or thought that it looked scruffy. The second explanation is its small size, and the 'creeping' style of walking that it has adopted. Both descriptions are quite true; this is the smallest of the banded penguins and probably the second-smallest penguin species, and it does have a rather tatty appearance in comparison to most other penguins. Like the other *Spheniscus* types, it has a black band that surrounds its chest, but rather than a nice clean, well-defined stripe, the plumage of this bird is rather broken and ragged. Around their head they do not have white chin plates, but a thin white line that runs from the beak to above the eye, curving around behind the cheeks and back to the neck and chest. Overall, it is quite distinct and it is difficult to mistake it for other species.

Like its nearest neighbour, the Humboldt penguin, one of the main challenges this tropical bird faces is the heat of the equatorial sun. The Galapagos Islands are volcanic and the rocky terrain is quite rough, with craggy black rocks. Where they can, these birds will find rock crevasses or volcanic caves to nest in, although they are also happy

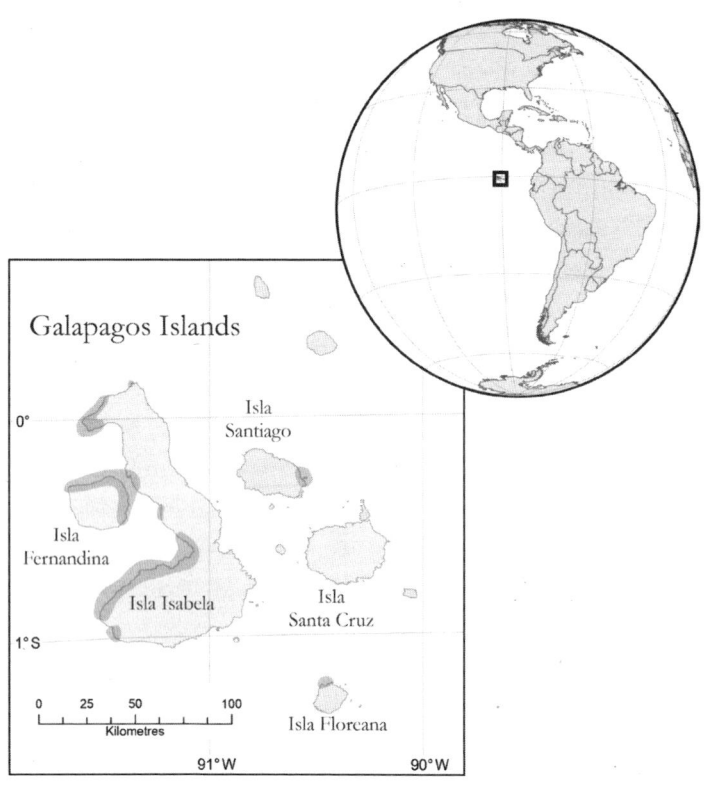

digging burrows if the rock or soil is soft enough. Their main aim is to find somewhere cool and shady. They often live in the vicinity of other types of seabird and line their nests with leaves, feathers and old bones to make comfy homes. When they are in the sun, they will try to cool down by panting or shading their feet and other non-feathered parts of their body with their wings. If there is a breeze, they will tend to hold their wings outstretched to get as much airflow over the exposed undersides of their flippers as possible. Like the other banded penguins, they have pink fleshy areas on their face, which they use to get rid of excess heat. Another strategy for keeping cool is to go out to sea for food in the daytime and only come back to their nests in the evening, or at night.

The Galapagos penguin eats small fish that abound around the coast when the cool Cromwell Current is strong. This nutrient-rich current is an offshoot of the larger Humboldt Current, which flows for thousands of miles northwards up the west coast of South America. As there are no true seasons on the equator, the birds can breed at any time of year and in good years they often raise two broods. They will not even try to breed in bad years, waiting it out until the good times (and the fish) return, hunting their prey at the surface and rarely travelling far from their breeding sites. The breeding range of the species is very restricted, limited to just a few rocky shores on the western side of the main island and one or two other small islets. The population is low and has fluctuated widely over the last few decades. This is a real boom-and-bust penguin, very dependent upon the fickle Cromwell Current. Like the larger Humboldt Current the strength of this ocean flow is ruled by the El Niño–La Niña cycle. This can have devastating effects on the penguin population. After the strong El Niño of 1982/83, the population fell to just 700 individuals, a drop of 70%, before recovering slightly in subsequent years. It is said that in those years there were no fish and many adults just starved to death. Other threats include introduced predators such as feral cats, dogs and rats, which can decimate individual breeding sites. And like many other seabirds, there is also a problem with fishing. Galapagos penguins have been known to drown in nets used by the expanding fishing industry around the islands.

Currently the population is thought to be around only 1,200 individuals, making this the rarest and one of the most endangered of all penguins. With so few individuals left and such a precarious dependence on ocean currents, conservationists are extremely worried about the survival of the Galapagos penguin.

Macaroni penguin
Eudyptes chrysolophus

Height: 70 cm, **Weight:** 5.5 kg
Range: Widely distributed throughout the sub-Antarctic
Population: 12.6 million, **Conservation Status:** Vulnerable
Diet: Krill, other crustaceans, fish and squid

The macaroni penguin is the most populous penguin in the world. It breeds over a wide range, on almost every sub-Antarctic island between South America and the Australian islands of Heard and Macquarie. There are even a few colonies in the Antarctic, although only in the most northerly parts on the South Orkney and South Shetland Islands. At least 258 breeding sites are known to exist, which hold over 12 million adult penguins. They are so numerous that it is thought that macaroni penguins eat more biomass in the oceans than any other bird, over 9 million tonnes each year. Exceptionally large aggregations live on several of the small, rocky fly-specks of islands that stick out of

the stormy, but bountiful, Southern Ocean. On the French islands of Crozet and Kerguelen, they breed at multiple sites in massive numbers at numerous colonies, with 2.2 and 1.8 million on each island respectively. However, even with these huge numbers, the future of the species is in trouble and populations have been in freefall in many places. At South Georgia, numbers were estimated at 10 million birds in the 1980s, but that fell to 5.5 million by the 1990s, and now numbers only around 2 million individuals. On the islands south of Australia the trend has been similar, but at their French strongholds the populations seem to be holding up. The IUCN Red List of threatened species lists them as 'vulnerable' due to their rapid population decline.

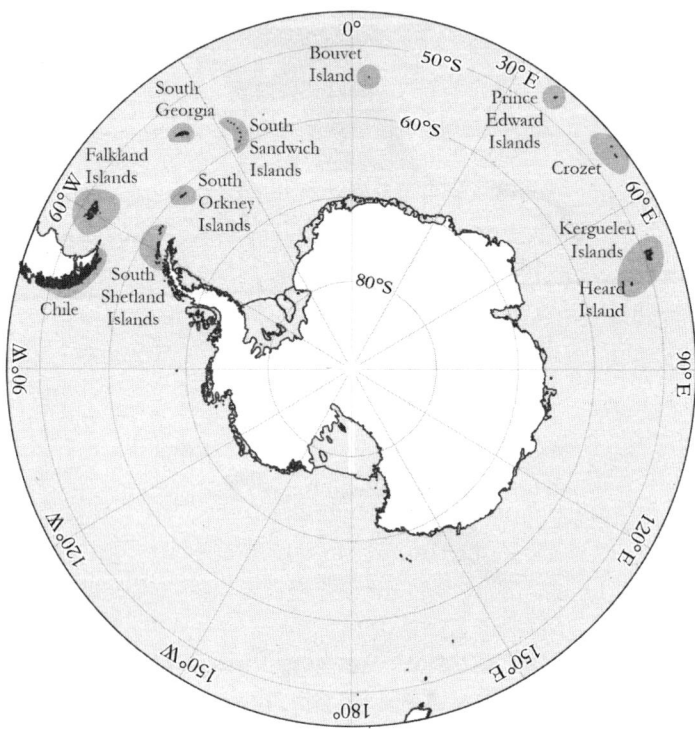

The name 'macaroni' has an interesting meaning. When British sailors first encountered the colourful birds, with their bright, flashy eyebrows, on the Falkland Islands in the 1700s, they named them macaroni, which was a name at the time for the fashionable fops and dandies in Georgian England. One of the few other references to macaroni (other than pasta) is in the song 'Yankee Doodle', written about the same time. If you remember the lyrics, it was written about a Yankee Doodle: a country bumpkin and a bit of a dandy who, by sticking a feather in his cap, became a macaroni, a sort of Georgian new-world hipster. Our macaronis may not be Yankees, but they are certainly dashing and charismatic.

Like all the crested penguins, this species of bird has long, bright yellow feathers above its eyes, something that biologists call an occipital crest. A similar coloured strip runs from the top of the beak and across each eyebrow as a yellow flash, and it then explodes into a tuft of bright, golden yellow feathers at the back of the head. Size-wise, it is probably only the outrageously attired northern rockhopper that has longer crest feathers. The rest of the penguin's body and plumage is a standard black and white.

As well as being the most numerous, it is also the largest of the crested penguins, standing on average around 70 cm tall and, as is the norm in penguins, the males are slightly larger than females. Weight can vary considerably depending upon the time of year. Macaronis will bulk up and be at their fattest just before their annual moult in the autumn, and be at their thinnest 4 weeks later, after the trauma of losing and renewing all their feathers. This process can change the weight of the bird by over 40%, from 6.4 kg to a skinny 3.7 kg.

Like most of the crested penguins, they love to breed in dense colonies, but they are not so house-proud or picky as the brush-tailed penguins. Nests can be made from small stones, but they can also be scrapes in the earth or mud, or amongst tussock grass. When breeding over such a wide range of habitats, it helps not to be too choosy. When they do nest out in the open, the nesting density can sometimes be as high as 1.4 penguin nests per square metre.

Northern rockhopper penguin
Eudyptes moseleyi

Height: 54 cm, **Weight:** 3 kg
Range: Mostly in the South Atlantic islands of Saint Helena and Tristan da Cunha, with a smaller number on French sub-Antarctic territories
Population: 410,000, **Conservation Status:** Endangered
Diet: Varies by location but includes krill, other crustaceans, fish and squid

What a hairstyle this little bird has. The northern rockhopper has the longest, most extravagant crest of any penguin. The long, thin feathers that make this crest stretch from above the adults' eyes to down as far as their necks really do make them look like they have long, straight, yellow hair. The shorter black feathers on the top of their heads also stand on end, making this a very extravagant bird indeed. The reason for such an elaborate crest is not really known but, as with many bird species, long and colourful plumage is often seen as a sign of a healthy individual and helps when choosing a mate. And this species goes to town when courting. It extravagantly shakes, rolls and quivers its golden locks when advertising to the opposite sex. Like some bleach-blonde 80s rocker in a mosh pit, it really knows how to let its hair down. The strange thing is, rockhoppers generally mate for life, so it only really matters having such a good hair display for a very short amount of time, when you first choose a mate. To me, it really doesn't seem to make that much sense having all that plumage and such a fine dress sense after that (although my wife might disagree!).

The rockhoppers are the smallest of the crested penguins, although the northern species is on average slightly taller than its southern relative, standing a diminutive 52–55 cm tall. As well as their yellow crests, they have bright orange or tile-red eyes, making their overall appearance even more striking. Like all crested penguins, they like to live on remote, windswept, rocky islands and use their fantastic 'rock hopping' ability to scale the cliffs here (more about this in the southern rockhopper section, on page 170). It is sometimes also called 'Moseley's' penguin after Henry Moseley, the naturalist on the *Challenger* expeditions that discovered and described the birds in the 1870s.

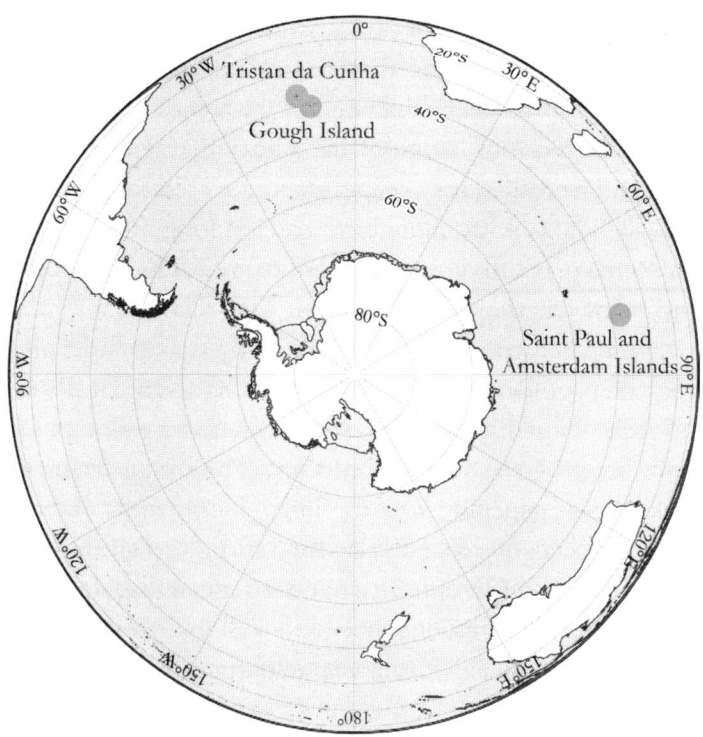

Around 90% of them breed on the UK overseas dependency of Saint Helena and Tristan da Cunha in the South Atlantic. The actual colonies are located on a number of small rocky islands, such as Nightingale and Gough Island, that surround the main island of Tristan da Cunha. The other 10% breed further south, on the French overseas islands of Amsterdam and Saint Paul. They live in different geographical and climatological regions from their cousins the southern rockhoppers, with the northern rockhoppers breeding and foraging in the relatively balmy waters north of the Polar Front – the convergence of warm and cold masses of Antarctic and subtropical waters that divides the Southern Ocean – while the southern rockhoppers always live in the chilly waters south of the front. This species is one of the most endangered of all penguins and the story of its recent past is not a happy tale. Populations are thought to have plummeted by 90% since the 1950s. Even before that, numbers were declining, mainly as the species was hunted for food, feathers and eggs, and used for fishing bait, ever since people discovered the islands. In more recent times these practices have generally been banned or are declining. More problematic in the last fifty years have been habitat destruction, overfishing and climate change, as well as predation by introduced species such as mice and rats. But the real killer has been pollution. In 2011, the cargo ship MS *Oliva* ran aground on Nightingale Island, spilling 1,500 tonnes of heavy fuel oil. It is thought that half of all the world's breeding population of northern rockhoppers may have been affected by the oil spill. Thousands were found on the beaches, and many were taken back to the main island for cleaning, but only a few hundred survived. It is impossible to say how many died, but the already fragile population was put on an even steeper slope towards extinction. The volume of shipping around the island is still increasing and fears remain that a similar incident could happen again.

It was estimated that in the mid-1800s over 2 million of these, the most flamboyant of all the penguins, existed. Today, only around 200,000 remain and numbers are still in steady decline.

Southern rockhopper penguin
Eudyptes chrysocome

Height: 40–50 cm, **Weight:** 2.5–4 kg
Range: Sub-Antarctic islands south of the Polar Front
Population: 2.5 million, **Conservation Status:** Vulnerable
Diet: Fish, crustaceans and squid

The southern rockhopper has a very similar body shape to its northern cousin. Perhaps if anything it is a little smaller, making it the tiniest of the crested penguins, averaging only around 45 cm tall. Up until 2009, the two species were considered the same, just called rockhoppers, with the Latin name *Eudyptes chrysocome, Eudyptes* meaning 'good diver'

and *chrysocome* derived from 'yellow-haired'. When the two species were split, the southern rockies got to keep the original name. And it is predominantly this yellow hair, actually the feathered crest above their eyes, by which you can instantly distinguish the two species. Whereas the northern brood have long, wild, flowing locks, often reaching down to their shoulders, the southern birds have a rather paltry offering. Southerners still have the golden yellow eyebrows, but rather than a mass of feathers flowing from the back of their heads they just have a couple of

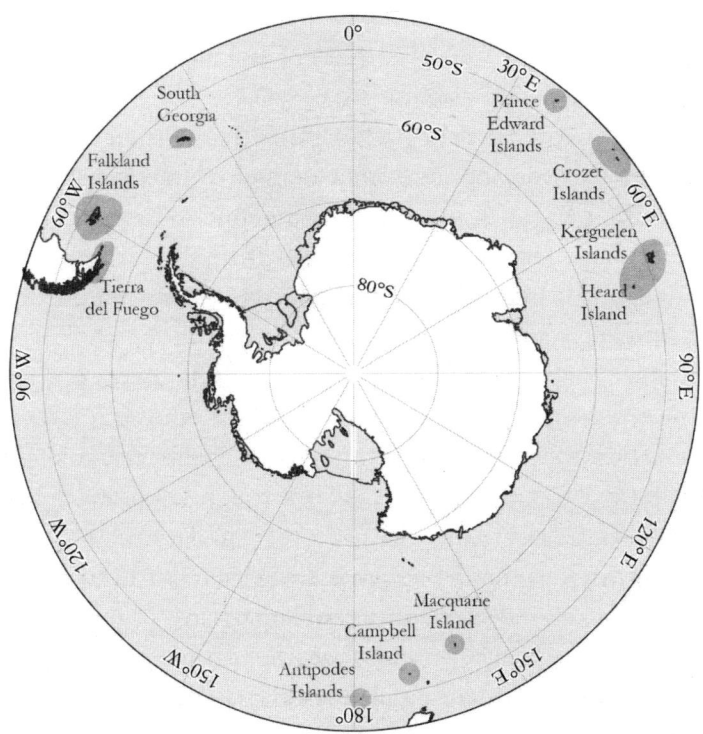

golden filaments stretching back beyond their eye line. So it seems to me a bit of a shame that the yellow-haired northerners didn't get to keep the term *chrysocome* and this went to its much plainer southern compatriot. Seeing pictures of the two together, it does rather puzzle you as to why they were ever classified as the same species.

Southern rockhoppers have a large range, breeding on sub-Antarctic islands across the Southern Ocean, pretty much all the way from Cape Horn, across the bottom of the Atlantic and Indian Oceans, to the Australian and New Zealand territories south of the Pacific. With such a wide range and with many colonies, their numbers, unsurprisingly, are much larger than those of northern rockies, outnumbering them by about five to one.

The name 'rockhopper' comes from their skill at jumping. You may think that penguins are clumsy, but rockhoppers buck this trend, being able to leap 6 ft in a single hop. That is about four times their body length. For context, the human world record for the same jump from a standing start is 12 ft, about twice our body length. Rockhoppers need this ability to leap between boulders and up the steep cliffs of the rocky islands in the Southern Ocean on which they live. They are tough beasts and have to cope with the tempests and waves of the roughest oceans in the world on a daily basis.

There is some debate about whether southern rockhoppers are one species, or made up of two or more (see the section Splitters and clumpers on page 23), with some scientists naming an eastern rockhopper version *Eudyptes filholi*. This is the portion of the population that breeds in the eastern area of their range around New Zealand. This penguin, traditionally called *tawaki piki toka* by the local Māori, is slightly different in appearance, having a white border around its beak and a hairstyle somewhere between the flamboyant northerners and the plainer southerners.

Like their northern cousins, the population of southern rockhoppers has been doing none too well. Numbers on some islands have crashed. On Campbell Island, a small sub-Antarctic island south of New Zealand, numbers have declined by 1.5 million, down 94% between the 1940s and 1990s. Other islands across the range have seen serious declines, with the overall population thought to be falling at a rate of 34% over three generations, above the metric used by the IUCN to trigger the 'Vulnerable' status that this penguin falls into. Reasons for the population declines are not exactly known, but warming oceans, driving their prey further south, and the introduction of non-native species such as rats on to many of the islands, are thought to be the most likely problems.

Snares penguin
Eudyptes robustus

Height: 45–70 cm, **Weight:** 2.5–4 kg
Range: Breeds only on Snares Island, a small island south of New Zealand
Population: 63,000, **Conservation Status:** Vulnerable
Diet: Krill, fish and squid

The Snares penguin is probably the hardest to visit of any penguin species. It lives on one very small group of islands, from which it takes its name. These remote, uninhabited, rocky fortresses are 200 km south of New Zealand, in the wind-lashed Southern Ocean. The whole land area of the island chain is less than 3.5 km^2 (a little over 1 square mile) and on that lives the whole global population of this bird. Even if you could get to such a bleak outpost of life, tourists are not permitted to land on the islands, which are a restricted nature reserve. So, catching a glimpse of these rare creatures is tricky.

In looks, the Snares is similar to other crested penguins, especially the closely related Fiordland penguin. The main difference is the bare patch around the base of its beak.

The crest is also slightly different. Starting near the beak and running above the eye and ending in short plume feathers, it is paler yellow and both shorter and thinner than other *Eudyptes* penguins. Its size and weight are pretty typical for its family and, genetically, it is pretty close to several other crested species.

On the islands, the birds nest under the scrubby trees or out in the open if they cannot find cover, often in groups of 50–500 nests, which for this species are scrapes of earth or small collections of twigs. They are fairly early breeders, turning up in August or September. They lay two eggs in late September and when they hatch the male usually looks after the youngsters for the first 3 weeks, while mum collects the food on a daily basis. They feed on krill and fish around the island, which has productive, rich waters in which to forage. Similar to other species, their main predators are giant petrels and skuas on land and leopard seals in the ocean.

Their small population in combination with their restricted breeding area is a concern for conservationists who worry that if some catastrophe hits the islands the whole species could be wiped out. That is why they restrict visits. Threats like bird flu or the introduction of non-native species would have rapid and dramatic consequences if they took hold. This species is one that really does have all its eggs in one basket (a 3.5 km^2 basket). That said, the population seems to be stable at the moment and the nests are counted by researchers once every few years. The main other concern going forward is climate change, which creates shifts in prey distribution and increases the chances of storms and bad weather.

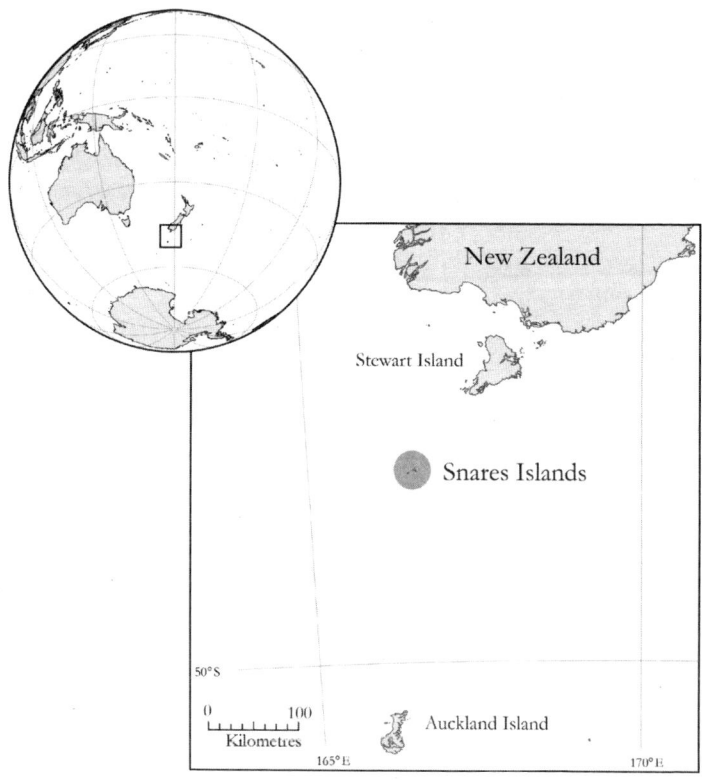

Many other rare avian species live on these small, isolated islands, including several species of albatross, and important populations of shearwaters and petrels. They also host two other endemic bird species found nowhere else, the Snares snape and the Snares tomtit. Not bad for one square mile of rock!

Erect-crested penguin
Eudyptes sclateri

Height: Up to 64 cm, **Weight:** 2.5–3 kg
Range: Breeds only on Antipodes and Bounty Islands, New Zealand
Population: 150,000, **Conservation Status:** Endangered
Diet: Krill, fish and squid

Erect-crested penguins live up to their name and are quite distinct from the other crested penguins. Their yellow crest feathers curve up from the nape of their beaks to above their eyes, ending in a gravity-defying hairstyle that points upwards almost vertically. It is like a double mohawk, which must take an ocean of hair gel to keep upright! They are the only penguin with an upward-pointing crest and those yellow feathers can be 6 cm above their heads – a tenth of their height. If northern rockhoppers are the long-haired rockers, then the erect-crested are the punks. Other than their huge hairy eyebrows, there is little difference between them and several other of the crested penguin species.

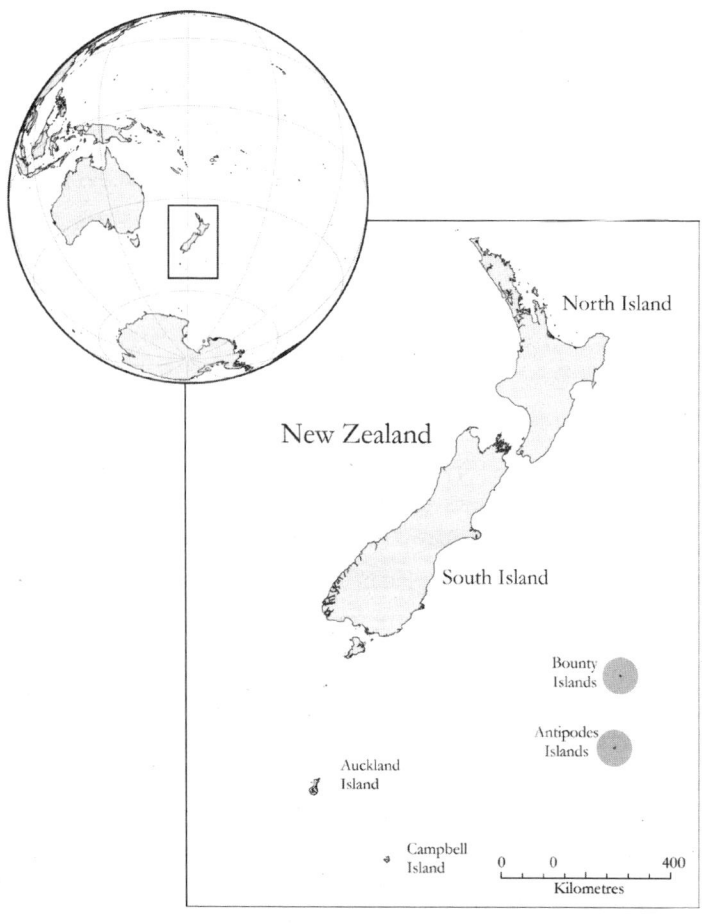

Like many of the *Eudyptes* family, the erect-crested penguin breeds in the rich waters below New Zealand, in this case on two small, isolated island groups called the Antipodes and Bounty Islands. The Antipodes are named for their geographic position, as they are almost at the exact opposite side of the globe to London. The Bounty Islands were named after the ship HMS *Bounty*, which discovered them just months before its famous mutiny. The thirteen tiny Bounty Islands have a combined area of only around 1.3 km^2 and are so covered in seabirds they have been made into a World Heritage site. The erect-crested penguins share these two island groups with southern rockhoppers and other seabirds, including the rare Salvin's albatross and Bounty shag.

These penguins have a very strange breeding habit; like many penguins they nest on rocky areas near the coast, but they lay extremely odd-sized eggs. Most penguins lay two eggs per clutch, with one often smaller than the other, but in this species the difference is severe, with the second egg almost twice the size of the first egg. It's the greatest difference in egg size of any bird. The first egg is usually discarded after the second is successfully laid, which seems a bizarre and slightly wasteful breeding strategy.

And their strange breeding strategy doesn't seem to be working. Like many species of penguins, their numbers seem to be falling at an alarming rate. Both sets of islands on which they live are exceptionally remote, so survey records are scarce, but the best available data suggests that numbers halved between the mid-1970s and 1990s and, although the decline has slowed since then, it is still worryingly high.

The exact reason for such a severe decline is not known. There are no non-native species on the islands, so the most likely culprit seems to be either changing ocean currents, which moves their food supply away from the islands, or interaction with the local fishing industry. Bycatch and net entanglement are known to be a threat to several other species of penguin, so it is quite possible that fishing may be the culprit for the erect-crested's decline. But there is so little data on this bird that the guilty party is hard to prove.

Fiordland penguin
Eudyptes pachyrhynchus

Height: 50-60cm, **Weight:** 2.5-3kg
Range: West and south coasts of New Zealand
Population: 12,500-50,000, **Conservation Status:** Near Threatened
Diet: Mainly squid and fish

If you look at the population status of the Fiordland penguin in the Birdlife International data list, it says between 12,500 and 50,000. That is a pretty big range and gives you some idea of the problems associated with studying this bird. The Fiordland penguin lives in extremely inaccessible parts of the New Zealand coast, mainly in the Fiordland District (yes, that is how you spell it; when Kiwi settlers originally named it after the Norwegian Fjords, they spelt it incorrectly!) from which it takes its name. This region has a complex coastline, which runs along the southwest coast of South Island. It is 1,000 km long, and, in the whole Fiordland area of 12,500 km^2, it has a population of just

SPECIES

a few thousand. The region is mountainous, forested and has virtually no roads, so access by boat is essentially the only option. A smaller portion of the population breeds in more accessible areas on Stewart Island, just off the south coast of mainland New Zealand, where they seem to prefer breeding in caves rather than under trees. It is in these easier-to-reach populations that much of the recent study on the species has been conducted.

The other problem in working out how many there are, or finding them at all, is that unlike other penguins they don't nest in colonies. Fiordland penguins are the antisocial cousins that like to be on their own. Usually, each pair finds some small, dark hollow under a bush or tree root to make their nest, a cosy cup of earth, lined with ferns and leaves.

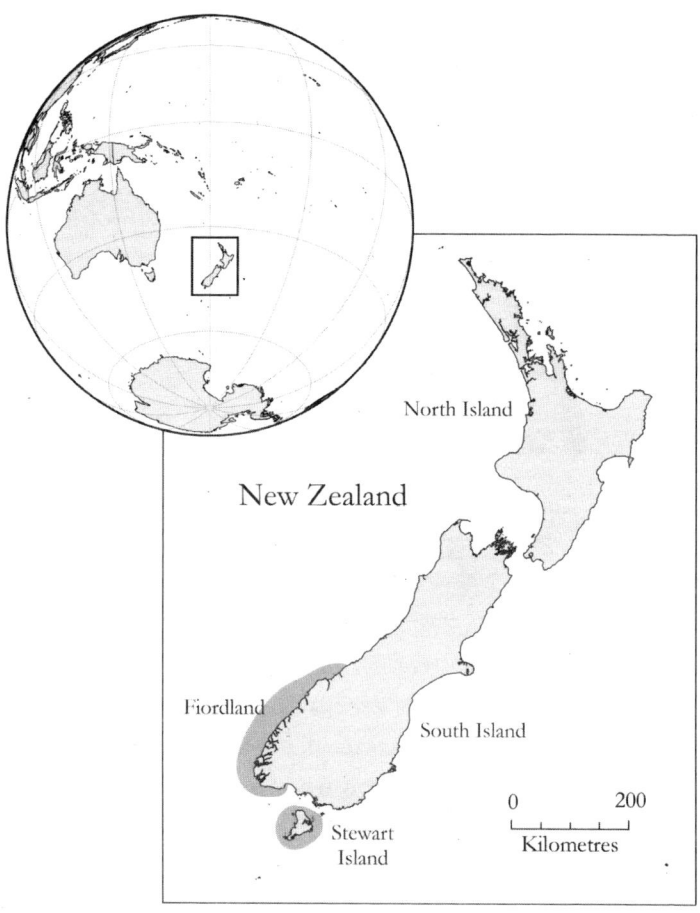

Other crested penguins that breed in noisy, smelly congregations of thousands are fairly easy to find, but the Fiordland is much more secretive.

In its native New Zealand, the Fiordland penguin is more commonly referred to as the Fiordland crested penguin or *Tawaki* in the native Māori tongue. The Māori name comes from ancient legends about a god called Tawaki, who came down to earth in human form, although exactly why a shy penguin should be named after this deity is a bit of a mystery. It looks similar to other crested penguins such as the Snares or macaroni, with a yellow line of feathers above its eye, which stretches to a crest towards the back of its head. Generally, these feathers are longer and slightly more yellow than the Snares', but not as long or as colourful as a macaroni's. The eyebrow stripe is much thicker, probably the thickest of any crested penguin, and it has 3–6 white stripes on each cheek that are only really visible when the penguin is angry or upset. Not that you are likely to get close enough to see them. I have watched them at a distance from a boat in Milford Sound, and as a tourist, or even a visiting researcher, that is pretty much as close as you are going to get.

As they are so difficult to find and study, we don't know as much about this secretive little penguin. They seem to forage close inshore and little is known about their diet, although some studies show that they eat a lot of squid. They have an unusual breeding strategy, breeding in the winter rather than in the spring or summer. Like most penguins, they lay two eggs, with the second egg much bigger. It is this larger egg that hatches first, and of the two chicks usually only the first survives. A grim but clever way of maximizing reproduction in a very changeable environment.

What we do know is that hundreds of years ago this species was much more common, living in and around the coast of much of New Zealand and even in southwest Australia. But the colonization of Māori peoples in New Zealand, who hunted them for food, and more modern colonizers in Australia, who brought many non-native animals with them, reduced its breeding range, so that now they hide away in the lush, remote, mountainous dark forest of South Island, away from humans. Non-native animals are still one of the major threats to the species, though. Dogs will attack and kill adult penguins, while stoats, rats and possums love to eat their eggs. One of the few native predators they have is the weka, a small indigenous flightless bird that is rather partial to penguin eggs.

With all the problems of surveying these elusive birds, working out how the populations are faring is tricky. A survey in 1990 reported that there were only 2,500 nests remaining, but more recent work on the more accessible nests around Stewart Island has suggested that the old survey grossly underestimated the true abundance, and more recent estimates advocate larger numbers. In truth, we really don't know how many there are, or how they are doing, but with increasing levels of threats, it really is something that we should work on.

Royal penguin
Eudyptes schlegeli

Height: 65–76 cm, **Weight:** 5.5 kg
Range: Macquarie Island, Australia
Population: 1.5 million, **Conservation Status:** Least Concern
Diet: Krill and other crustaceans, fish and squid

The royal penguin is the only penguin with a white face. Apart from that, it looks very much like a macaroni and

SPECIES

genetically it is quite similar, so there is quite a lot of debate amongst scientists as to whether it is a separate species, or just a regional morph of its much more widespread cousin. In the wild, royals and macaronis have been known to interbreed, although this is a rare occurrence. The white face makes royal penguins instantly recognizable, so even if genetically very similar to macaronis, it is likely that this distinctive little bird will remain designated as a separate species. The title 'royal' is thought to have originated from the yellow crest on its head, which looks like a crown, and the specific name *shlegeli* is in honour of the German zoologist Hermann Schlegel, a counterpart of Charles Darwin who strongly opposed Darwin's theory of evolution, which has a certain irony when you realize that genetics might be used to prove that it isn't really a separate species at all.

They live only on Macquarie Island, mainly on the mainland of the island with a few on a small group of rocks called Bishop and Clerke Islands just 37 km to the south. Macquarie Island is a sub-Antarctic island, located about halfway between New Zealand and Antarctica, but politically it is part of Tasmania, Australia.

The behaviour of the royal penguin is very similar to that of the macaroni. It breeds on beaches and rocky areas near the shoreline, making a small scrape in the ground or

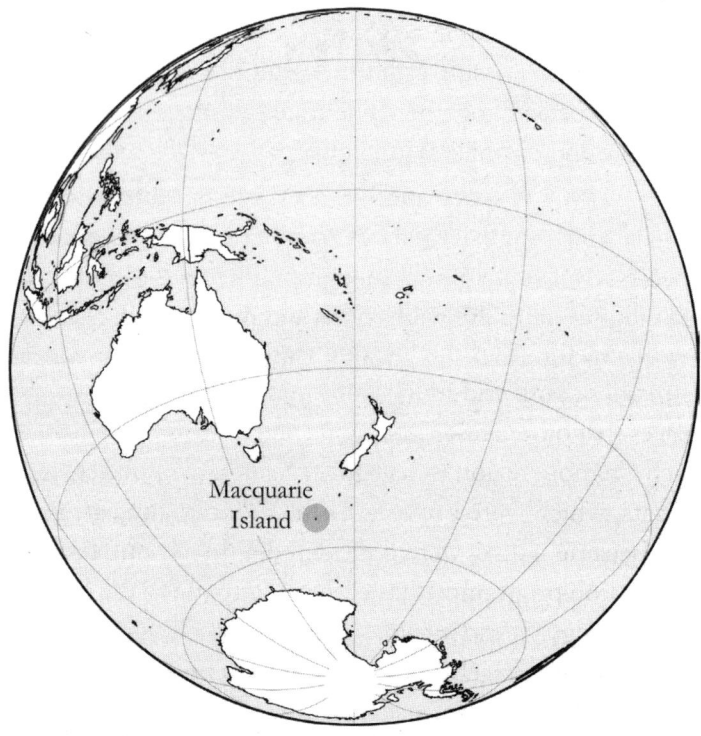

between the rocks. The birds are fairly shallow divers, only occasionally diving below 100 m, and feeding on much the same as the other crested penguins – krill, fish and squid. They lay two eggs, but, like several other penguin species, usually only one survives. Often the first egg, which is usually slightly smaller, is abandoned immediately after or sometimes just before the second egg is laid. Overall, the first egg has about half the chance of being kept, and even less chance of being successfully reared, as the second.

Although the royal's breeding range is very limited, what it lacks in distribution it makes up for in numbers. On this small island, measuring just 34 km by 5.5 km, there are 570 breeding colonies, some of which are huge. By far the largest site, at Hurd Point, has about 1 million penguins packed together. The total population is thought to be around 1.5 million individuals and, unlike many other types of penguin, numbers appear to have been fairly stable over the last few decades. The main threats affecting other penguin species, like habitat loss and conflict with the fishing industry, do not seem to be impacting the royals on Macquarie. This may be due to geography as the island is further south and much more remote than most of the places that other crested penguins breed, and fishing here in such a remote region is negligible. Non-native species, such as cats, which were brought to the island many years ago and became native, were considered a threat, but a recent eradication programme has eliminated them, to the benefit of the penguins and lots of other indigenous seabirds. The impacts of climate change do not seem to be affecting them, either. This may be because they live further south, closer to the Polar Front with access to other productive seas around Antarctica.

The island is a state reserve and, apart from a few scientists, uninhabited. Though it wasn't always like that. A hundred years ago, there was a booming industry in penguin oil on Macquarie and over the spread of a few decades at the end of the 19th and beginning of the 20th centuries, millions of royal penguins were slaughtered for their oil (see Chapter 5). Due to public protest (helped by a number of polar celebrities), the killing of penguins for oil was banned in 1934 and the industry was shut down. The penguin numbers have slowly recovered, but it is unknown if they have ever reached the levels of population that existed before the butchery.

Yellow-eyed penguin
Megadyptes antipodes

Height: 62–79 cm, **Weight:** 3–8.5 kg
Range: East and south coast of South Island, New Zealand, and the sub-Antarctic islands of Auckland Islands and Campbell Island
Population: 2,600–3,000, **Conservation Status:** Endangered
Diet: Mainly fish, with some squid

The yellow-eyed penguin is unique amongst penguins in having not only yellow eyes, but an almost totally yellow head. It does not have a crest, and in stature it is bigger than any crested penguin, but still much smaller than kings or emperors. It is the only member of the *Megadyptes* family currently in existence. The genus name *Megadyptes* means

SPECIES

'large diver' in Latin, and the species name *antipodes* refers to the place where it lives – the antipodes, or opposite end of the earth to Europe. The colouration on its head is interesting. It has a thick, bright, pure yellow stripe that runs from the back of its eye around the back of its head to link up with its other eye, and although most of the rest of the head is also yellow, that colour is not pure or bright. For most of the head, the yellow feathers are paler and mixed in with a few black plumes, giving a more speckled-greyish appearance, especially on the cheeks and on top of the head. The eye, from which it takes its name, is orangey yellow with a black pupil. It has a tall, slim appearance in comparison to most other smaller species.

Although it is now the only member of its own penguin

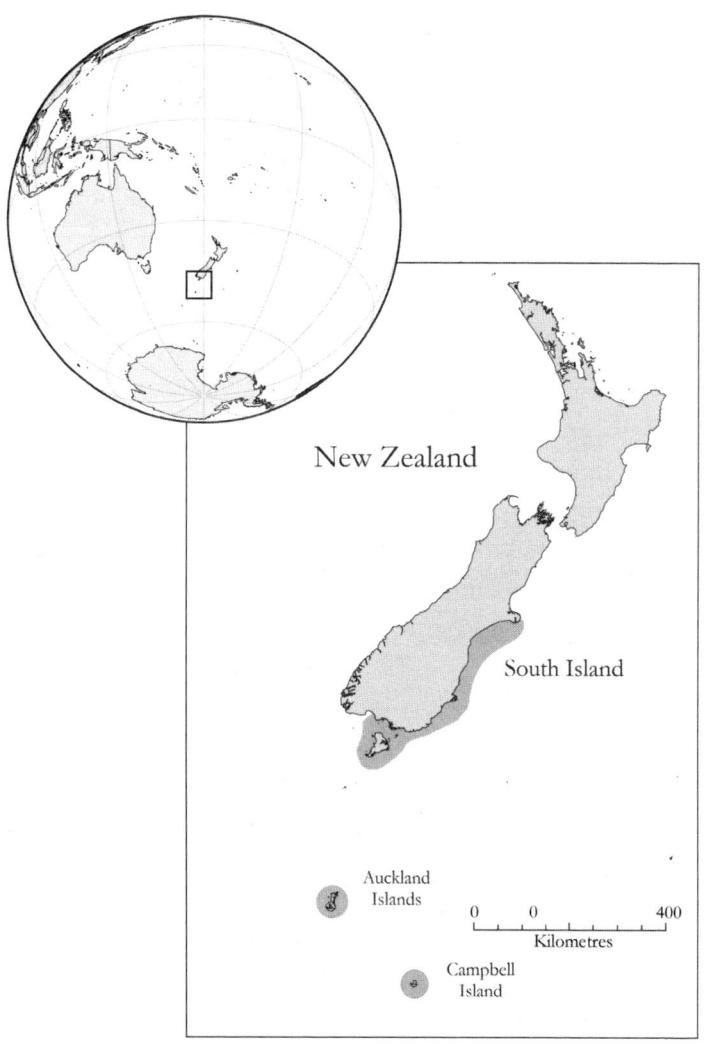

family, recent research has revealed that it was not always alone. Until about 500 years ago, a slightly larger similar penguin, *Megadyptes waitaha*, also existed on mainland New

Zealand, breeding up the east coast of South Island and on the more southerly parts of North Island. But, with the colonization of the Māoris, numbers plummeted, until the *waitaha* eventually became extinct due to overhunting around 1500 AD. At that time, the yellow-eyed penguin did not breed on mainland New Zealand, but the lack of competition for resources saw them colonize parts of South Island after the extinction of its cousin. The yellow-eyed have been hunted by the Māori too. They are called *hoiho* by the indigenous people of the islands, a name meaning 'noise shouter', in reference to their squawking call, but this smaller immigrant seemed to be able to cope with the level of hunting that its extinct predecessor could not. Perhaps because they like to nest in dense bush away from the coast, and like the Fiordland crested penguin often don't breed in large colonies but away from each other, they are less easy to find and catch.

Today, there seem to be two distinct populations of yellow-eyed penguins. One on South Island, New Zealand, and some of the nearby coastal islands, with another population much further south in the sub-Antarctic realms of Auckland and Campbell Islands. Neither population is large. The total number of individuals including both populations is probably below 3,000, making this one of the rarest penguins on earth. At breeding sites on South Island, numbers have fallen significantly in the recent past. Records from surveys in 2019 showed less than half of the breeding population compared to 2012. It is thought that the main reason for this dramatic fall in the adults inhabiting this area was from fishing bycatch. The use of gill nets was identified as the likely culprit. Gill nets can be lethal to penguins, as the birds try to snatch fish caught in the nets and get trapped themselves and drown. Gill-net fishing is now banned in New Zealand waters within 6.4 km of the coast, but the birds, although not great travellers, do tend to forage well outside this limit, so numbers are still likely to be falling. With such a small starting number any adult mortality can have significant impacts on the population.

Other threats, such as predation from non-native dogs, stoats and ferrets, also pose a problem on the mainland, but less so on the offshore islands. Further south, in the Southern Ocean on Auckland and Campbell Islands, the threats of overfishing and alien species are less acute, but the remoteness of these islands means that monitoring is not as regular as on mainland New Zealand. The true picture is that we really do not know how many of these unique penguins are left. It seems as though, if things do not change, and as the Māori did to their sister species, newer, more industrial colonists might enact that same fate on the yellow-eyed penguin.

Little penguin
Eudyptula minor

Height: 41–45 cm, **Weight:** 1 kg
Range: Southern coast of Australia and coasts of New Zealand
Population: 470,000, **Conservation Status:** Stable
Diet: Small fish, squid and krill

The little penguin is known by several names. As well as 'little' it is also called the fairy penguin (in parts of Australia); the blue penguin (the name usually used in New Zealand); or the little blue and the *Kororā* (by the native Māori). It lives up to two of its three names: it is the smallest penguin species and the only one that has blue plumage, but scientists who work with it say that it is rather aggressive and its demeanour is not fairy-like at all! Its diminutive height, rarely over 45 cm, is exaggerated by its stooping posture, as unlike other penguins it doesn't tend to stand up straight. It is the only one of the *Eudyptula* genus and is quite different in looks and behaviour from most other penguins. *Eudyptula* means 'good little diver', although conversely it probably dives shallower than any other species, although it is excellent at hunting in the coastal waters it frequents. It doesn't like to travel too far, either. Some species will go hundreds of kilometres to get a crop full of fish, but this penguin is a home bird and will only go out to a maximum of 10–20 km from its breeding sites.

As mentioned, they have a dark blue or blue-grey coat of feathers on their back and a white front. The transition between the white and blue can be indistinct, making their plumage look a little tatty, compared to some of the other better-dressed species that have nice, sharply defined

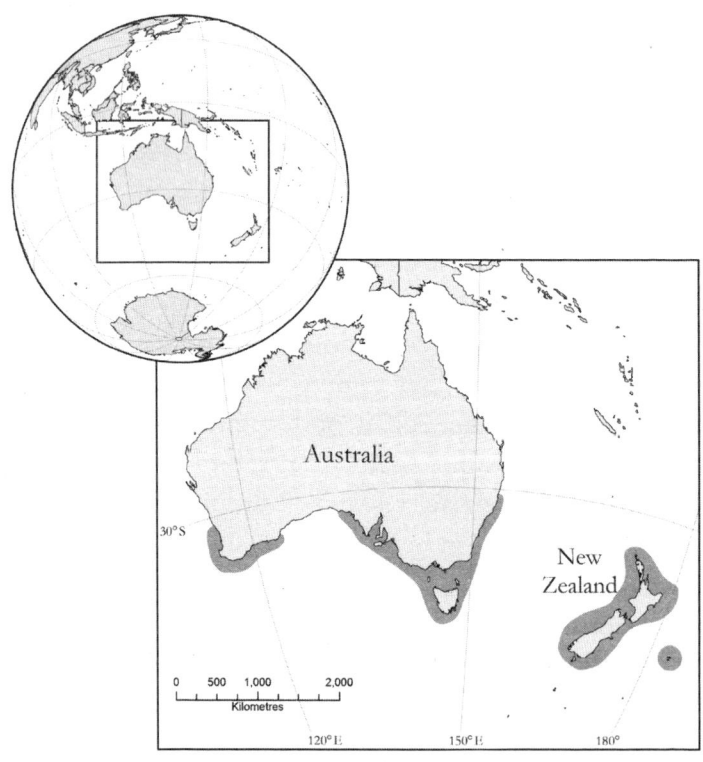

patterns. They have grey eyes and black or dark grey beaks to match. Unlike other species, little penguins are quasi-nocturnal, leaving the nest before sunrise and only coming back on to land after dusk. But they are not truly nocturnal, as in the daylight they are not sleeping but foraging out in the ocean, using the light of the sun to hunt, and travelling the short distance between the water and their nests in the safety of darkness.

Little penguins breed in shallow burrows or caves and crevices, although some Eastern-Australian birds build nests of small stones. Their colonies are not densely populated, often having a good distance between nests and, although they are quite nervous around humans, they will often nest under houses and wooden structures, as they find these easy to dig under. The adult birds stay at the breeding site all year round and, in more productive areas, they can have two broods per annum. Due to their small size, the incubation and chick-rearing cycle is relatively short in comparison to their larger cousins.

Little penguins have a range of distinctive calls and noises from grunts, brays and yaps to prolonged trilling calls and, interestingly, different regional groups have quite different calls. As they breed in fairly populated areas they have been relatively well researched over the years and many landmark studies on behaviour have been conducted on this species.

In Australia they live on the southern coast and around Tasmania and, in New Zealand, around almost all the coastline. However, there are noticeable differences between the penguins in each of those regions. Genetically, this species is probably the most diverse of any penguin, with variations in breeding season, plumage colour, calls and nest preferences. It is thought that the little penguins in western Australia and southeast Australia are different subspecies, with the western ones generally bigger and breeding two months earlier than the eastern birds. New Zealand birds are different again, mainly in their behaviour. But it is a specific subspecies called the white flippered penguin that really gets the Kiwis annoyed. Pretty much all New Zealanders, and many other biologists, think that it should be classified as a whole different species. As its name suggests, the white flippered penguin (*Eudyptula albosignata,* relating to those white markings on the flippers) has white edges to its flippers, which other little penguins do not have. It is also much less blue in colour, having a dark grey back, plus it is heavier and slightly taller. It only breeds on the east coast of South Island in New Zealand and has quite a limited range and population of only around 2,000–4,000 breeding pairs. This would make it endangered if it was classed as a separate species, especially as numbers are declining.

Taken overall, with all the different subspecies, the little penguins are considered to have a stable population and are classified as 'least concern' by the IUCN, but that hides a range of threats and concerns. Some scientists think that many populations are under severe pressure. These birds often breed near urban areas and have suffered from habitat loss and disturbance from coastal development across their range. Additionally, introduced species such as cats, dogs and ferrets have had a very negative effect in some areas, especially near human habitation. A single predator can decimate a small colony, especially when the birds, which evolved with no mammal predators, have no defence against the invaders. Fishing is also thought to be impacting the foraging birds, mainly through the accidental bycatch of the birds themselves. Finally, climate change is also starting to have an effect. Warming of the oceans changes currents and prey availability, leading to poor breeding in warmer years. In Australia, the recent warming climate has led to a number of extreme heatwaves. These hot spells result in overheating and death through heatstroke, especially for the penguins that breed in the open, rather than in burrows or caves.

But there is still some good news for this, the smallest of our feathered friends. People care about penguins, and conservation efforts at many sites have reversed the declines. Erecting fencing has kept out predators, and if human disturbance and habitat loss have been an issue, many penguins have been rehomed, with cosy nest boxes in protected sites. This is often paid for by tourism. Living so close to humans means that seeing this species is easier than most and the little penguin is one of the most visited penguins worldwide. As they will often come ashore together around dusk in large numbers, watching their daily commute home has become a popular attraction. The money raised goes back into conservation, which seems to be halting the population decline at many breeding sites. Whether the local conservation efforts will be enough to offset the effect of future climate change, well . . . we will have to wait and see.

King penguin
Aptenodytes patagonicus

Height: Up to 100 cm, **Weight:** 9–18kg
Range: Sub-Antarctic islands, 45° to 55° South
Population: 1.1 million, **Conservation Status:** Least Concern
Diet: Mainly fish, some squid and octopus

The king penguin is probably the best dressed, brightest and most colourful of the penguin species. The deep, solid orange of its cheek patches tapers down to its neck and blends into the white at the top of its chest. On their backs, they are a smart slivery blue-grey rather than the black of most other penguins, with a sharp black line separating it from their white and pale golden chest. It stands at 1 metre tall, the second largest of any penguin and thinner than most. Its beak is the longest beak of any species, measuring in at up to 14 cm (nearly 6 inches) long, and even this is colourful, with a bright pinky-orange plate running along its side.

Regal indeed, and boy do they know it. They take great pride in their looks. All seabird species need to preen, clean and oil their feathery coats to ensure that they are waterproof. I have watched many types of penguin and it seems to me that the king spends more time looking after

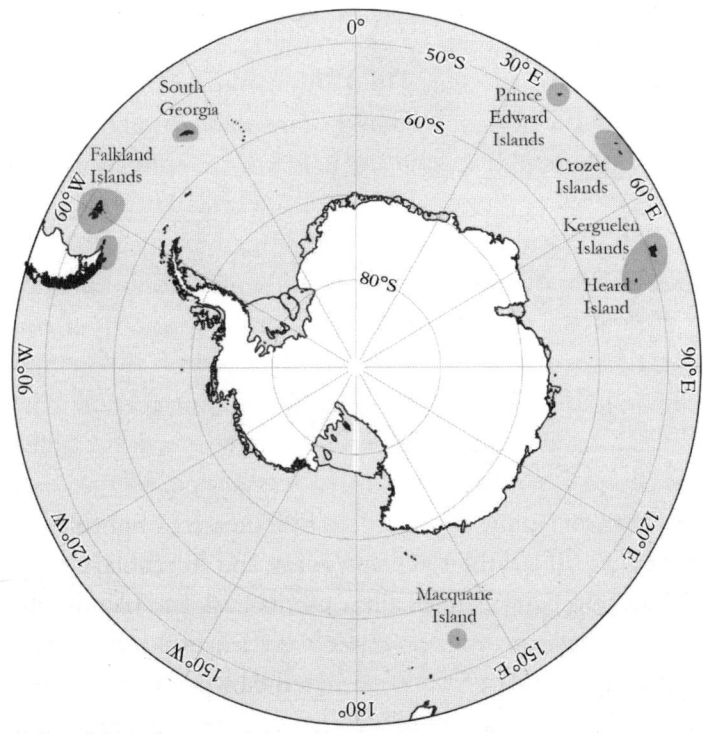

its plumage than any other species. It can be quite comic watching them contort their extremely flexible necks, rubbing on their waxy oil from a gland just above their tails and trying to reach every part of their body.

The king penguin is the second-largest penguin, only slightly smaller than the only other great penguin, the emperor. The two are fairly similar at first glance, especially if you don't see them together, which you never will, as the king never goes into the seas around Antarctica and the emperor never travels away from the ice surrounding the Antarctic continent. The king is somewhat slimmer than its imperial cousin, weighing in at around 15 kg on average, although this weight can vary greatly depending upon the time of year. The Latin name for this genus is *Aptenodytes,* which has a combined meaning of 'good diver' and 'featherless'. The king and emperor are both very good divers, but they are certainly not featherless! This reference probably comes from the fact that their plumage was so fine and the plumes so dense that some people thought that they had fur rather than feathers. The specific Latin name for the species is *patagonicus*, given to it by the first person to describe the birds, the British explorer and illustrator John Frederick Miller, in the early 1700s. He called the penguins *patagonicus* as he first found them off the coast of Patagonia in the South Atlantic. Today, there are still one or two small colonies off this coast, but in reality almost all the population breed on a number of the larger sub-Antarctic islands in the South Atlantic and Southern Indian Oceans. There are two recognized sub-populations; the South Atlantic one is based mainly on South Georgia, with a few smaller colonies on the Falkland Islands, while the Indian Ocean subspecies, recently named *Aptenodytes patagonicus halli*, lives on the South African-owned Prince Edward Islands, the French Southern Territories of Kerguelen and Crozet, as well as the Australian Heard, McDonald and Macquarie Islands. The largest populations are on South Georgia, Kerguelen and the Crozet Islands, each of which has over 300,000 breeding pairs. All these islands are close to the super-productive waters around the Polar Front, the oceanographic boundary where the cold waters around the Antarctic meet the warmer subtropical waters to the north. The birds utilize this nutrient-rich region to provide food for their chicks. Kings never venture too far south, and you will never see them amongst the sea ice.

These birds can dive to great depths, usually over 100 m and occasionally over 300 m, mostly in search of small fish (*Myctophids*), which they chase down in the chilly deep waters. Their particular favourites are lantern fish, a small, plentiful fish that has bioluminescent patches along its sides which glow in the dark. It is thought that this feature helps camouflage them near the surface, but it also makes them easy prey for a fast hunter in the pitch blackness of the ocean depths. Did I mention that this fish is plentiful? Many scientists believe that lantern fish are the most numerous vertebrate species on earth and have an estimated biomass of 600 million tonnes, swimming around in the mid layers of the earth's oceans. Kings are amazingly fast swimmers, chasing down their prey and snapping up the fish easily, the bioluminescence making them easy to find. To see their prey in the depths, they have evolved the most amazing eyes, specially adapted to focusing on both land and water and, although their eyes look fairly small to us, beneath the skin they are huge, taking up over half of the skull cavity. Both eyes together make up an area larger than their brain! King penguins can take over 400 lantern fish in a single week-long foraging trip, before bringing them back to their single chick in a specially adapted crop in their throat.

They like to breed in huge colonies, on or close to the beach, often with other penguins or close to fur and elephant seals, which compete for the best real estate on the beaches. If you are lucky enough to visit one of the really big colonies, like St Andrews Bay on South Georgia, you will be assaulted by a cacophony of noise and the sights and smells from hundreds of elephant seals, thousands of fur seals and hundreds of thousands of penguins, along with all the predators and scavengers that depend on such a rich concentration of life.

King penguin colonies are somewhat different from those of the smaller penguins. For a start, they are less monogamous than other species. Most penguins go back to the same partner each year, but that only happens about 25% of the time with kings. Also, they only ever lay one egg, and they do not build a nest. Instead, they balance their eggs on the adults' feet. The parents take turns in looking after the egg until it hatches, almost 8 weeks after laying. Unlike the emperor, the kings don't move much during this period, each having their own spot on the beach. They are fiercely territorial over this spot, but, even so, the breeding density of their colonies is often still high, sometimes reaching over two penguins per square metre. We have already documented the long, strange and rather extreme upbringing of the king penguin chick, as it is left over the winter by its parents, starving in these snow-bound islands, before mum and dad return in the spring (see page 66). To cope with the winter cold and the months without food, the chick can weigh as much as an adult at the start of the winter period, before losing half its bodyweight prior to its parents' return. It's thought that this is the only vertebrate species where the young actively lose a larger percentage of their bodyweight halfway through their development. Protection from the cold comes from a layer of fluffy brown down feathers, which makes an excellent insulating layer against the snow and biting winds. This layer has to keep them warm for 10–13 months before the young chick fledges into its much smarter adult plumage. For the first few weeks of life the adults will constantly guard the chick, but at around 6 weeks old the chicks are large enough to fend off the skuas and giant petrels that predate younger birds and they become semi-independent. After this, both adults go out to forage for food to feed up the hungry youngster before the onset of winter. At this stage, the colony splits into separate areas of chicks and adults. By now, the fluffy brown chicks are large and quite different-looking birds compared to their parents. Over their long and traumatic winter, chick mortality is quite high for the young kings, often much higher than for most other penguins. You may think that these birds are awful parents, but it is how they have evolved and most chicks survive each year. However, even in the best-case scenario, an adult pair will only raise two chicks every three years, so populations of kings are slow to recover if anything befalls them.

So many large birds packed into such a small area will always attract predators and each colony will support a resident population of leopard seals, skuas and other hunters and scavengers. Rubbing shoulders with fur seals and elephant seals on the packed beaches can be hazardous, especially as the big bull fur seals are sometimes quite partial to a penguin supper.

The king penguins' relationship with humans has not always been happy. Today they are one of the most loved and recognizable of penguins, found in zoos, on chocolate wrappers and in countless adverts. But go back a few decades and people were exploiting these birds in an altogether darker way. They have been killed for their feathers, their eggs have been stolen, they have been eaten and squashed for their oil. Originally, sealers would kill them for meat and for oil for their stoves, but later they were then targeted on a more industrial scale for their oil (see Chapter 5). Populations on Macquarie Island, where the worst of the industrial exploitation happened, almost went extinct, but since the cessation of oiling in the early 20th century, the numbers have slowly recovered back to their original levels.

Today it is thought that populations overall are stable or increasing, one of the few penguins that seems to be bucking the trend. But scientists still worry, especially about climate change. Almost all of the colonies are located where they are because of their proximity to the productive Polar Front. Research has shown that, with warming oceans, this front is slowly moving south, away from the islands, potentially making foraging trips longer and the breeding locations less suitable.

SPECIES

Emperor penguin
Aptenodytes forsteri

Height: Up to 120 cm, **Weight:** 30–45 kg
Range: Around the coast of continental Antarctica
Population: 512,000 adults
Conservation Status: Near Threatened (but under review)
Diet: Fish, krill and squid

Introducing his imperial majesty, the emperor penguin. It's hard for me when writing about emperor penguins not to break into a rash of superlatives: it's the tallest, heaviest, most extreme penguin; it can dive deeper than any other bird; and it has a crazy breeding strategy that means that it has to face some of the coldest environments on earth. To do this, it has a whole host of bizarre adaptations, often unique to this amazing animal (see Chapters 3 and 4). It is a bird that I have studied for fifteen years. I've tracked them by satellite, counted them by drone, put GPS tags on them, collected guano samples and watched as they have fallen foul of warming temperatures around their Antarctic home. During this time I have come to love this fantastic creature. I said at the beginning of this book that scientists should not really have a favourite type of animal. Well, penguins are my favourite and emperors are very much at the top of the list.

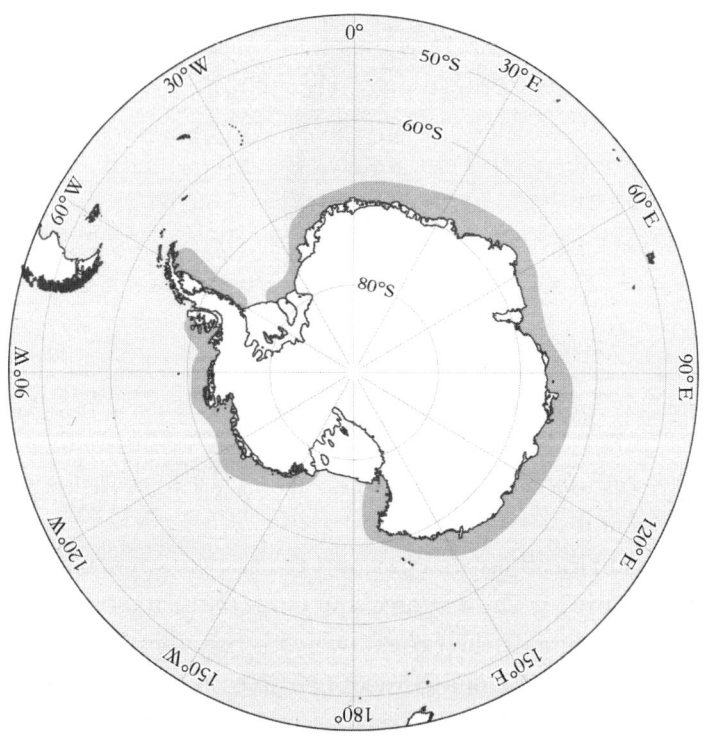

SPECIES

They say that emperors need to be big to withstand the extreme cold and keep insulated in the place they call home. A smaller bird just would not have enough body mass to survive breeding around the coast of the frozen continent. An emperor can stand as high as 1.2 m tall (that's 4 ft in old money), although in reality it rarely seems to reach this tall, as much of the time you see the adults hunkered down conserving warmth in the frozen wilderness. It's only when they stand up high and stretch up their necks, which, similarly to king penguins, are extremely flexible, that you really see their true height. The adults will do this when looking around or communicating with other penguins. Unlike all the other species that raise their heads to the heavens when they call, the emperors usually call with their beaks pointing at the ground, trumpeting a deep baritone sound unique to each individual. They are heavy too, weighing as much as 45 kg (that's a hundredweight of penguin). I know: I've tried picking them up to weigh them. To give you a reference, that's as much as a thirteen-year-old child, twice as heavy as any other penguin.

They are less colourful than the only other member of the *Aptenodytes* genus, the king penguin, although in photos they often look quite similar and images of them are regularly confused in the media. The main tell-tale characteristic is that the emperors' yellow cheek patches are much less bold than the kings' and fade to white as they spread down to their breast. The emperor's patches are open and merge into the white of their chest, which is often tinged a very pale yellow. If you saw them side by side, it would be easy to spot the difference, as the emperor is obviously much taller and stouter. But you will never see these two species next to each other, as the ice-bound emperor doesn't venture much beyond the realm of the pack ice and certainly not as far as the sub-Antarctic islands where kings breed. Occasionally there are reports of juvenile emperors turning up on sub-Antarctic beaches and recent tracking studies have shown that young birds do often travel much further north than adults, sometimes going as far as the Polar Front. When I once asked a behavioural biologist about this, he likened them to teenagers, with a lust to see the world, before coming back home to start a family.

Being heavy and often rather bulky, they are not the most agile of penguins. They are the only species that don't porpoise when swimming. Or at least they don't manage to leap fully out of the water. They try, often breaking the surface when near the ice edge, but they have too much mass to get airborne, unlike their smaller cousins. They are also quite slow on land. In fact, everything seems slow and considered for the emperor; it's like they are the Ents of the penguin world. They never run. They walk with a slow, rocking shuffle, wings held tightly against the body, swaying from side to side and gripping the slippery ice with formidable claws. If they need to go faster, they will flop on their bellies and scoot off, tobogganing across the ice at a surprising speed, driven by those strong talons.

They breed around the coast of Antarctica, where fast ice forms for long enough for them to complete their breeding cycle. Fast ice, or land-fast sea ice as it is sometimes known, is ice which forms when the sea freezes and is stabilized by being attached to a static object, usually land.

The term 'fast' in this context relates to 'stuck fast'. It's a bit of a contradiction as this fast ice does not move and is therefore the slowest and least dynamic type of sea ice. Exactly what constitutes land in Antarctica, where the fast ice may attach itself to, is a confusing topic. Land in Antarctic is often water – frozen water, in the shape of the edge of the ice sheet, or the cliffs of floating ice shelves or glaciers. Occasionally, rocky outcrops exist which fast ice binds to, but that is rare around the coasts of continental Antarctica. Around the western Antarctic Peninsula there is a lot of rock outcrop at the coastline as its temperatures are warmer, but it's too warm for stable fast ice to form for long enough for the emperors to complete their breeding season. So, most of the colonies are found around the more southerly ice-bound coasts of the continent.

Being large, our imperial friend needs an extended period to raise its chick to adulthood and, like the king, it cannot complete this cycle within the space of the short austral summer. But whereas the king leaves its chick over the winter on some windswept sub-Antarctic island, the emperor does not have that choice. In these climes, it is far too cold to leave its offspring over the winter, so it takes the rather extreme strategy of breeding in the winter. It is the only animal in Antarctica to do this. It is worth reiterating some of the details of that strange and brutal breeding cycle, as it is key to many of the other facts about this bird.

Emperors turn up at their breeding sites in late March or early April. In the southern hemisphere seasons are reversed, so this is their autumn. At this time of year, the sea is starting to freeze over and the adults will choose a place that has solid ice and not too far a walk from the edge of it. The males will turn up first, followed a week or two later by the females.

The places where they choose to breed are interesting. As we have mentioned, they need stable sea ice, and this exists around much of the coast of the continent, so they can pick and choose just the right spot. The Adélie penguin, the only other penguin that breeds in these locations, has to find a rocky outcrop to make its nest. But there is often no rock along some stretches of the coastline for thousands of miles, so the Adélies' breeding geography is rather restricted to a limited number of locations. This is not a problem for the emperors and in many regions they have the place to themselves. When you look at a recent map of the distribution of colonies you will see that the locations (sixty-six known colonies at the last count) are strung out like a necklace around the continent, each almost equidistant from each other, about 250 km apart, which equates to the maximum foraging range during the breeding season.

Like all penguins, they will go through a fairly complex courting ritual, although this species is rarely monogamous, the least faithful of any penguin, and almost always finds a new partner each year. Whether this is due to the fact that they have no fixed nests, so it's difficult to find their old partner, or whether it's because they are in a hurry to get on with it while the ice is stable, we don't know, although my bet is on the former, as emperors rarely seem to be in hurry to do anything.

Once the courting and mating have finished there are a few weeks of calm before the female lays her egg. Only ever one single egg. She gives birth on to her feet and then passes the precious cargo on to the feet of the male bird, where he keeps it warm in a special pouch-like recess at the bottom of his torso, covered by a flap of skin. This happens around mid-May time, during the late autumn (in the southern hemisphere the seasons are reversed), before the worst of the really cold weather sets in and before the sea ice has reached its maximum extent. The females then leave the breeding site to feed in the Southern Ocean and regain their strength. Over the next 8 to 10 weeks, between May and August, the male emperor will protect the egg. This is the period of greatest cold, when the thermometer drops below -50°C and howling hurricane-force winds called katabatics race down from the continental interior in speeds in excess of 161 km an hour, driving the wind chill levels to astronomic extremes. To conserve warmth the remaining males form a huddle. They pack as close together as they can, backs to the wind, heads down, letting those amazing feathers do their job. Scientists have used sensors to measure the temperature in the centre of those huddles and were astounded to find that they can get as high as 25°C. This is far too warm for our well-insulated bird; they overheat in the centre, so when the males in the middle get too hot, they move to the edge. Meanwhile, those at the edge with their backs to the wind will be getting cold, so they will barge or wriggle their way towards the centre to warm up. This means that the huddle is continually moving, a boiling mass of penguins, each penguin readjusting to keep the perfect temperature.

As the birds on the windward side tend to get colder

faster, they move to the centre more often, so over time the whole huddle slowly meanders downwind. We can see this in the satellite images: large huddles of penguins leaving behind a winding brown snail-trail of poo, depending on the direction of the wind.

Eventually the egg will hatch and, if all goes well, the female will return just as the chick emerges, to give it its first feed. It will take 8 to 10 weeks for the egg to hatch and by the end of the process the male will have been on the ice without food for four months. During this time it will have lost up to 40% of its body mass. At this time of the season, in August, the sea ice is reaching near its maximum extent, so mum may have had to walk huge distances, sometimes in the order of 50 km, over the frozen landscape to find her family. For the first few weeks after it is born, the chick is too small and scrawny to keep itself warm, so it will stay on the adults' feet and the adults will take it in turns to feed it. As the youngster gets older, it will start to become more independent and will venture out more often. After a couple of months, the chick will be big enough to withstand the cold and then both parents will go off to find food, leaving the growing chicks together in crèches. These groups are sometimes looked after by older, experienced adults, maybe a parent that has lost its egg. But in my experience, at the colonies I've been to, the chicks will stay put or wander around their group without venturing too far away, otherwise their parents will not be able to find them when they come back. When mum or dad does come back, they will find their chicks by calling out to them. The chicks each have a unique fluttering call, which they will continually chant in the hope that the next meal is soon on its way back. The adults seem to be able to recognize their chicks by this call alone and don't have the same problems of fending off pretenders as the smaller species. No silly chick chase for an emperor – far too undignified! By early December, the chicks, which will now be three-quarters grown, will begin to get their adult feathers. These will form underneath their fluffy grey coats and as they grow will push out the downy feathers. They will then be waterproof and able to go into the water. At this stage, they will assume juvenile status and soon have to fend for themselves. The parents will give them their last feed and then leave the colony. The adults need to put on weight for their moulting period in a couple of months' time, so they are eager to leave. Eventually, hunger will drive the youngsters to also abandon the colony, take to the water and leave the ice behind. They will not return to the site for four or five years, when it will be their turn to become parents themselves. By the time they fledge in late December, if all has gone well, the extent of the sea ice will have diminished and the ice edge will be much closer to the colony, so that the now almost fully grown chicks do not have far to go to reach the water.

Like all penguins, the adults have to moult and change their old feathers annually and they do this in February, the latter part of the southern summer, while standing on the sea ice. They travel to find the very last of the pack ice where it's still concentrated in places like the eastern Ross Sea or western Weddell Sea, often swimming many hundreds of kilometres to reach the right spot. Four weeks later, the emperor will have new clothes and be in a rush to catch fish, squid and krill, diving down to those crazy deep depths to bulk up enough to go back on to the ice and start their extreme breeding cycle once again.

Emperors breed almost as far away from humans as it is possible to get. They were the last penguin species to be discovered and most of their colonies have never been visited by people on the ground. They have never been hunted, and they face no pressure from fishing, as no boats can penetrate the pack ice that is their icy home, and there is no pollution or any invasive species to worry about. So, you would assume that they are safe. But you'd be wrong. Emperor penguin numbers are declining at a worrying rate, and it's because of us. Human-induced climate change is warming the planet, and even the coldest place on earth is starting to feel its touch.

The fast ice on which the emperors breed is a fickle breeding platform, thin and fragile at the best of times, and a slight change in sea temperature can lead to it forming too late or breaking up too early. If it breaks up too early, before the chicks fledge into their waterproof feathers, they will be immersed in the frigid waters of the Southern Ocean and die of exposure or drowning. This is starting to happen already, at an alarming rate. In 2022, an exceptionally early break-up event led to nineteen out of the total of sixty-six emperor penguin colonies having ice-break-up before the chicks had fledged fully. Tens of thousands of chicks perished. With sea ice continuing to decline, it looks like this is a story that will only get worse in the coming years. Recent models of the population trajectory suggest that the species will be almost extinct by 2100. Actual data shows that the situation is even worse and numbers are declining at an even faster rate. The difficulty is that this is not something that can be fixed on a local level; you cannot put the ice back. Only stopping global warming will halt the decline, and there is not long to go. It is ironic that this amazing bird, loved by millions and which, in most cases, will never see a human and lives further away from people than any other species, is being driven towards extinction by our actions.

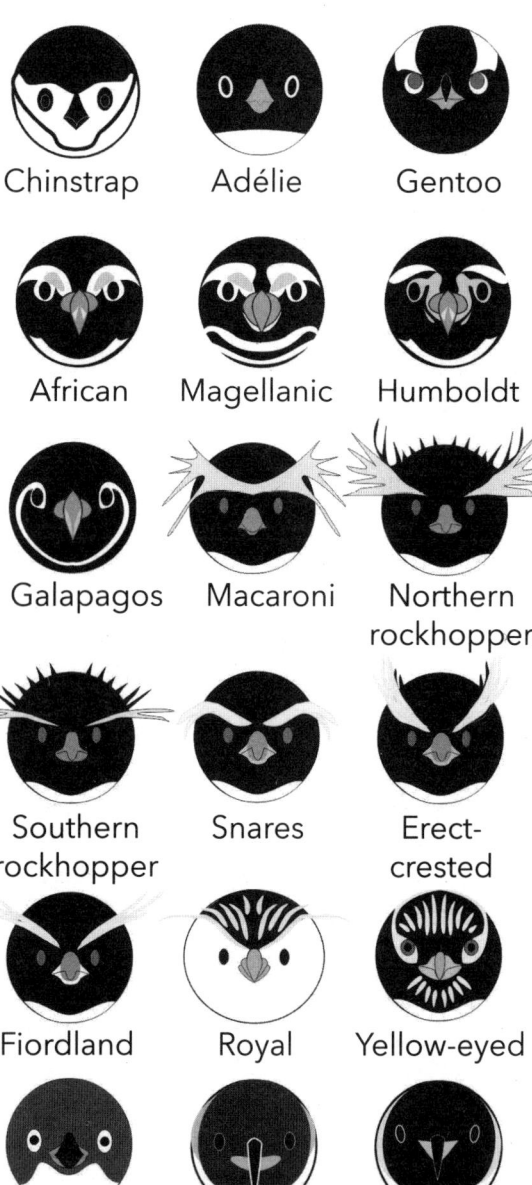

Chapter 7:
INTO AN UNCERTAIN FUTURE

Threats to penguins

Unfortunately, penguins are vulnerable to many human activities. Even when they are not being directly exploited, indirect mortality and breeding failure from habitat loss, overfishing, introduced species and climate change pose a serious threat to many of the eighteen species of penguin. Most penguins only have one, or occasionally two, chicks per year, so their population growth is slow. When something bad happens, it can take a colony decades to recover. They always breed on the coast, in areas that are often ripe for human development, and they tend to forage in areas of high marine productivity, which are prime fishing grounds. They are slow to adapt and are very loyal to their native breeding grounds, so they will often keep returning to a site, even when it becomes unsustainable. Like many seabirds that breed in large, dense colonies, diseases like avian flu could be catastrophic. Perhaps even more worrying in the long term, many species are already threatened or will be adversely affected by the growing spectre of climate change.

Habitat destruction

Imagine the perfect penguin residence. A nice ocean headland, not too steep, with plenty of flattish ground to nest on, with a good beach to come ashore. Sounds good? In temperate locations, such as South Africa, Australia, New Zealand and South America, these areas that have often been settled by penguins are also prime locations for human development. Humans and penguins don't mix very easily. Over the last few decades, many areas around penguin colonies have been built up, and as urban growth has accelerated, towns have mushroomed, and industrial development has encroached on to beachfront habitat. Roads along the coast create barriers or hazards, especially for the smaller penguin species which often come ashore late in the day or after sunset, making them difficult for drivers to see. Building can disturb nests or force penguins away from colonies to seek quieter surroundings. Some penguins have adapted to these new developments. In Australia and New Zealand, beachside residences are often built of wood and raised above the ground. Little penguins will often build their homes under the floorboards of wooden cottages, but they are noisy neighbours and having them under your house can be a smelly business. Yellow-eyed penguins, which also live in New Zealand, prefer cool, shady woodlands to make their nests, but a large percentage of their native forest habitat has been cleared for agriculture. Even when yellow-eyed penguins are allowed to breed on agricultural land, their nests are in danger of being trampled by livestock; they become easy prey for predators; and, with the lack of shade, they can suffer from overheating in the sun. Today, few colonies remain on the mainland, with most of the population restricted to offshore islands. Their populations have been in serious decline for decades and, despite continued conservation efforts on South Island to protect them, the last census data showed only just over a hundred breeding on the New Zealand mainland. Today, for this species that was once common, only a few thousand individuals remain.

Pollution

For penguins, pollution problems come in two main forms: oil spills and plastics. Human development around the coast often brings marine traffic and with it the possibility of oil spills. Penguins spend most of their time at the sea surface and when exposed to oil in the water the consequences are often fatal. Even a minor oil spill can be catastrophic, and there are dozens of examples in the last few decades in temperate regions like South Africa, Australia and South America of thousands of penguins and other seabirds succumbing to marine oil spills. A number of important penguin breeding grounds, such as those of African penguins around the Cape of Good Hope, are on some of the busiest shipping routes in the world. Oil spills, either clandestine or accidental, are a constant threat and major conservation efforts are needed to monitor ships and mitigate any incidents. In 1968, a major oil spill in South Africa killed at least 15,000 penguins. The outcry was so great that a new organization was set up, the South African National Foundation for the Conservation of Coastal Birds (SANCCOB). This initiative provides a 24/7 rescue service for sick and injured seabirds and abandoned seabird chicks, responding to oil spills all along the South African coastline. Since that time, it has answered the call to many disasters, cleaning and saving thousands of birds. South Africa and other governments have attempted to bring in more stringent laws to avoid spills, but, with ever-increasing numbers of ships, incidents continue.

Even in Antarctica, which we often think of as pristine, there have been a number of incidents. Many research stations are near large penguin colonies, especially on the Antarctic Peninsula. So, most of the resupply and research ship traffic passes close to, or through, the feeding grounds of huge numbers of Adélies, chinstraps and gentoos. It is around the Peninsula that tour ships also ply their trade. Each year, around a hundred large cruise ships go back and forth several times a season to the area, most targeting penguin colonies as their prime destination. Any leaks or spills of oil in these environments can be extremely damaging, especially because, in the very low temperatures of the Southern Ocean, the biological processes that break down oil-based contaminants are extremely slow, so oil persists in the natural environment for much longer. With the potential threat to Antarctic wildlife so high, in 2011 the International Maritime Organization banned the use of the worst types of fuel, heavy fuel oil, in Antarctica. Heavy fuel oil is particularly bad, as it has many toxic impurities and even the smoke from burning it can cause health problems. It is thick, viscous and, if spilt, extremely difficult to remove. The ban, for all ships below 60 degrees south, was the first of its type, anywhere in the world.

The second major pollution threat to penguins is from plastics. In all oceans, plastic pollution is becoming an increasingly severe problem. Waste plastic can take hundreds of years to decompose and, even then, it breaks down into microplastics that can be ingested by fish and other marine organisms. Higher predators, like penguins, are particularly susceptible as they eat the fish, inadvertently consuming the microplastics which their bodies cannot get rid of, a process called 'biomagnification'. Over time, levels of contaminants build up until they become toxic. Some of the highest levels of microplastic found in penguins have been in Galapagos birds feeding in the Pacific Ocean.

Even in the clear seas of Antarctica, microplastics can be found in the guts of penguins. Field analysis has identified that one in five gentoo penguin droppings contains microplastics, and other research shows that the level of plastic pollution is getting worse. As the levels of plastic waste in our oceans increase, it is likely that all penguins and all other marine predators in all oceans will be affected by plastic pollution.

Alien invaders

Aliens have been killing penguins for years. No, not the bug-eyed monsters from outer space. In science, 'aliens' is a term used by biologists to identify non-native species, those animals that have been brought, usually by humans, to otherwise pristine environments. The repercussions of introducing these alien invaders can be devastating. Often, penguins and many other birds that nest with them have evolved on remote islands away from land-based predators, so the introduction of alien species can, and often has, led to the decimation of whole colonies and areas being abandoned. Today, many of the main threats to penguins are from non-native species.

This problem is not new. Early visitors to South America and the sub-Antarctic islands brought animals with them, sometimes for food or as pets. Sealers and whalers introduced pigs and reindeer as a ready food source to many islands, as well as cats and dogs. Over time, pigs, rabbits and reindeer became wild, cats and dogs became feral, and the native wildlife became prey. Many of these remote islands, which had never known terrestrial mammals, suddenly had huge problems. The vegetation was eaten or trampled by grazing reindeer, and the birds were predated by fierce new hunters that the original inhabitants had no defence against. In South America and South Africa, packs of feral dogs caused havoc at many penguin colonies on the mainland and drove several species offshore, to nest on small islands where the hounds couldn't reach them. This problem has been especially serious for another subtropical species, the Humboldt penguin, which was driven from its island homes by the mining of guano, only to be predated by dogs on the mainland. Conservation efforts to stop this have included constructing fences and barriers around colonies to try to keep the dogs at bay.

But these larger animals were not the worst offenders. What made the greatest impact were the animals that the explorers and whalers brought in by accident: the rats and mice that all old wooden ships had aboard. These smaller creatures overran many islands that were strongholds of penguins. The huge congregations of birds on South Georgia, Kerguelen and Crozier all suffered, especially in the areas around the old whaling stations. Rats would eat the eggs and mice would nibble away at the chicks until they died. Even the adults were not immune, as they had no understanding of or defence against the tiny invaders. Penguins can be quite feisty and got off relatively lightly with the mice attacks compared to some other bird species. It was the albatross that had some of the greatest problems. These serene animals, some with the greatest wingspan of any bird, are peaceful by nature and had no answer to the incessant nibbling of the mice. On some southern islands, like Tristan da Cunha, mice developed a thirst for blood, biting and sucking the blood from albatross chicks until they expired. For albatross and penguins alike, population numbers on the larger islands around human habitation fell dramatically.

On several of the sub-Antarctic islands there have been recent concerted efforts to kill the alien invaders. Rats and mice have been eradicated with huge, island-wide rat-poisoning campaigns. Tonnes and tonnes of poisoned bait have been dropped by helicopters over large swathes of the islands and the reindeer, originally brought to places like South Georgia as meat for hungry whalers, have all been shot. The mice have been more difficult to get rid of, but success has been achieved on a few of the islands. On some of the main sub-Antarctic penguin breeding grounds, such as South Georgia and Macquarie Island, and on many small islands of the Falklands archipelago, rats and other aliens have been eliminated completely.

Rat

In New Zealand, the introduction of mammals has been catastrophic for many of the native birds. Unlike almost any other large landmass on earth, New Zealand has no native land mammals (except for a couple of species of bat). This strange evolutionary quirk occurred as the islands broke away from the other landmasses far back in geological time, before mammals evolved, and the deep cold channel between Australia and New Zealand has protected them from mammal incursions ever since (see page 19). Only the birds could get there, so, over time, the two temperate islands evolved a host of unique indigenous bird species that filled all the evolutionary niches that mammals did on other continents. When humans arrived and brought mammals, the effects were disastrous. The birds had no defence. Each year, rats, stoats, ferrets, pigs and dogs can kill up to 60% of chicks at some colonies. In many areas, the penguins that nest on the mainland like the little penguin and yellow-eyed penguin have suffered ongoing population decline. Conservation measures to control predators are ongoing at many sites around the country, but the battle is a difficult one and only time will tell if the penguin populations can be saved.

Fishing

Fishing brings a number of threats, but the most severe are bycatch, where birds are caught in the fishing nets by accident, and overfishing.

Many of the fish that our little feathered friends prefer are also harvested by humans, often on an industrial level. As penguin colonies are frequently large, they need to be located at the sites of highest marine productivity, which are also the most intensely targeted by the fishing industry. For many species, overfishing is currently the greatest cause of population decline. In South America, along the coast of Peru and Chile, the Humboldt penguin is particularly affected. It relies on the colossal quantities of anchovies that are brought close to the coast on the rich, cold Humboldt Current, from which the penguin gets its name. But fishing boats also depend on this resource, landing almost 1 million tonnes of anchovies per year from this area alone. The quantity of fish is very variable; in the El Niño years, when fish stock plummets, the fisheries catch what they can, leaving very little for the penguins. Famine amongst penguins is commonplace, not just in the chicks but in adults too. A severe El Niño in the 1982/83 season reduced the population from approximately 20,000 to around 5,500 individuals, as a result of reproductive failure and starvation of adults, a phenomenon exacerbated by fishing.

Other penguins suffer too. The African penguin is in real danger from fisheries. Its preferred prey is the tasty sardine, also a target for industrial fishing fleets in the area. Unregulated levels of overfishing along the coast have left penguin populations in freefall. Like the Humboldt penguin, whose numbers were once depleted by guano mining, the African penguin population numbers fell dramatically over a hundred years ago, due to historical egg collecting. With today's threat from overfishing, numbers that were once counted in millions are now reduced to around 20,000 pairs, a fall of 98%.

Further south, around the remote cold waters of the Southern Ocean, fisheries are also taking hold. Here, the penguins mostly feed on krill, the small, shrimp-like crustaceans that sustain the food chain of the region. To say that krill are numerous is an understatement. It is thought that the biomass of krill in the Southern Ocean is around 400 million tonnes, the largest biomass of any animal in the world. As each krill weighs on average about 0.5–1 gram, it that means there are around 4–8 hundred trillion individuals. This vast bounty is the main menu for almost every predator that lives here: fish, seals, whales, seabirds and, of course, penguins. But over the last fifty years, people have also started to fish for krill. In the 1970s and early 1980s, as whale hunting became uneconomic, fleets, mainly from the Soviet Union, turned from catching whales to fishing for krill. By this time the industry was harvesting up to half a million tonnes of krill a year. It could have been worse; the large baleen whales that used to be the main predator of krill around Antarctica had been hunted almost to extinction, so, for penguins at least, the competition for krill had been reduced and in some areas it was thought that penguin numbers had risen due to the surplus. However, concerns about potential increases in the krill catch were one of the main drivers of the establishment of CCAMLR in 1982. The full name, the Commission for the Conservation of Antarctic Marine Living Resources, is a bit of a mouthful, so we will just call it CCAMLR, like everyone else does. This international conservation body was set up to manage fisheries in the Southern Ocean, setting quotas for krill and fish all around Antarctica, with the objective of preserving marine life and only allowing sustainable fishing that will not impact either the environment or krill and fish stocks. The commission's work is challenging; as it is based on consensus between all members, and getting things changed and implementing new initiatives can be slow and difficult. Overall, since its creation, it has been effective in avoiding the levels of overfishing that much of the rest of the world has experienced, in the rich cold waters of Antarctica.

The second major problem with fishing, that goes hand in hand with overfishing, is bycatch. This is the term used when wildlife is caught by accident. The various types of net and line fishing can cause serious mortality. Fourteen of the eighteen species of penguin have been recorded as bycatch by the fishing industry. All the temperate species are badly affected. Little blue and yellow-eyed penguins have been particularly badly hit. In Tasmania a single fishing net has been recorded to have killed fifty little blues in one haul. The worst type of net is the huge drifting gill net; these nets can stretch 60 km long and float vertically through the oceans like impenetrable walls of ultra-thin, nearly invisible nylon wire. They are designed to catch larger fish around their gills, but that size also happens to coincide with catching penguins behind their flippers, making it impossible for them to escape. As well as little penguins, the situation for Humboldt, Magellanic and yellow-eyed penguins is just as bad, with bycatch thought to be one of the main reasons for their continuing population declines, even in the areas where land-based conservation has secured their breeding habitat. In some areas around New Zealand, gill nets have been banned close to penguin colonies, but even other types of fishing can cause problems. Trawling in Argentina and Uruguay is thought to be affecting Magellanic penguins. At the moment the populations of this bird are still reasonably large, and, although many birds are caught, governments believe that it is not yet a critical pressure on the populations. The state of the population for Humboldt penguins is not so rosy, with overfishing, historical exploitation and introduced predators driving their numbers down. For this species, bycatch by the fishing industry could be the straw that breaks the camel's back and drives them towards extinction.

Climate change

One threat that is very difficult to mitigate, on a local level, is climate change. The warming of land, sea and atmosphere due to the burning of fossil fuels has had profound and ongoing consequences for many animals, and some species of penguin have been particularly badly affected.

One impact is on food availability. Warming oceans affect the distribution of the penguin's prey. We have seen that for the South American species, like the Galapagos and Humboldt penguins, the cold currents are key to their survival. It is predicted that with climate change these currents may falter, or the periods of warm waters may last longer, which would have devastating consequences.

On the sub-Antarctic islands, penguins breed in huge numbers. It is believed that the geographic location of many of these islands, near the productive Polar Front in the Southern Ocean, is the key reason for the abundance of life there. But with warming oceans, some scientists predict that the Polar Front will migrate south, towards Antarctica and away from South Georgia, Crozet, Kerguelen and the other bastions of penguin productivity. Penguins are central-place foragers, which means that they are tied to a single geographical location and have to go out on their daily or weekly commute from that place. As there are very few islands on which to breed in the Southern Ocean, these rocky outposts are prime real estate, often the home of millions of birds. But if ocean currents change, the fish and krill, which are not bound to any particular geographical location, will move with it.

This will make the foraging trips of the seabirds that nest on the islands longer and more difficult. Scientists worry that this will drive the populations of many species down and may lead to local extinctions. It is thought that this has started to happen and that several islands are already being affected. But teasing apart the slow, insidious effects of climate change from the other drivers of penguin population dynamics in the region is difficult. Two other major drivers may be pushing penguin populations down on the sub-Antarctic islands. One is the recovery of whales after the decimation of their populations in the 20th century: whale numbers are increasing rapidly and this may be affecting food supplies. The other is commercial fishing around many penguin colonies, especially South Georgia and the Antarctic Peninsula. These fisheries are some of the best regulated on the planet, and quotas are set each year, but if you cannot say for certain what is affecting the population trajectory – climate change, whale recovery or fishing – it is difficult to set those quotas effectively.

And it is not just in the cool waters of the Southern Ocean that warming waters are becoming a problem. Off the hot coasts of western Australia, marine heatwaves create changes in the distribution of the small fish that little penguins rely on. Researchers believe that this is the main reason for the population decline that they have seen in the region.

However, it is in Antarctica where the consequences of a warming planet have the most serious implications for penguins. As the sea warms, it freezes less easily and the sea ice that surrounds the continent forms later in the autumn and breaks up earlier in the summer. This has implications for two penguin species, the Adélie and the emperor, but it is the emperor that has the most to lose. The emperor penguin uses the frozen sea as a breeding platform, and without it, it cannot raise its chicks. It needs at least 9 months to complete its breeding cycle, from courting and choosing a mate in April, staying on the ice throughout the whole of the Antarctic winter, to when the chick fledges into its adult plumage in December.

So, the adult emperors choose their breeding locations carefully, with sea ice that is just right. But as conditions change, many of these colony sites are becoming unsuitable. If the summer season is warmer than usual and sea ice does not form in April, the breeding season is delayed. This will mean that the chicks, which hatch later, will not fledge until January, by which time the sea ice might have already broken up. With the warming waters and stormier weather the sea ice also breaks up earlier in the year, before the normal fledging season in December. Recently, we have seen many examples of the ice breaking up well before December, sometimes in October or November, months before the emperor chicks are fully grown.

In 2022, warm conditions southwest of the Antarctic Peninsula in the Bellingshausen Sea led to early sea ice break-up at four of the five emperor colonies there, and over the whole continent, that year, almost a third of all colonies suffered breeding failures due to poor ice conditions. When the sea ice breaks up early, the chicks will go into the water without their sleek black waterproof feathers. The grey downy feathers that cover their young bodies are great at protecting them against the snow and freezing winds, but they are not waterproof and become easily waterlogged.

Emperor chick

As their flippers are not yet developed, they cannot swim or jump out of the water, so falling into the ocean at this stage is a death sentence for the little chicks. It is thought that when a colony has early sea ice break-up, virtually none of the chicks survive. If this happens regularly, as it is starting to do at many colonies around the continent, populations will start to decline as generations of youngsters are lost and do not recruit back into the adult population. Predictions suggest that emperor penguins, probably the best-known and most celebrated of all penguins, may be extinct before the end of the century.

Unfortunately, climate change is not something that can be solved at a local level. You cannot put the ice back or refreeze the ocean. People have asked me if you could build a platform that emperors could breed on, but it is not that simple. Even if you could engineer something, transporting it to where the birds breed would be very difficult, plus it would be highly unlikely to withstand constant bashing by million-tonne icebergs, and there is no guarantee that the penguins would even use it. What's more, penguins eat snow to rehydrate, so if there is no snow they cannot keep hydrated. In short, emperor penguins are adapted to the sea ice and if there is no sea ice there will be no emperors. It is as simple as that. We just have to hope that humanity can solve the problem of global warming before it is too late for this amazing animal.

Penguin conservation

As I am writing this book, the International Union for Conservation of Nature, which runs the Red List of endangered species and has a group specially devoted to penguins, lists the penguin family as the second most-endangered seabird group. Overall, two thirds of the species are under some level of threat, with a third at the endangered (the second-worst) status and one, the African penguin, listed as 'critically endangered', the most severe status before extinction in the wild. It is telling that most of the Antarctic species, with the exception of the emperor, are of 'least concern', while the ones that are most concerning are those that live closest to humans. The population status of penguins has never looked so grim.

But the popularity of penguins and their often-precarious conservation status has meant that numerous people and organizations are trying to help save these endangered animals. All over the world, there are many charities and conservation bodies doing fantastic work to help reverse the decline. In many cases, this work is having a positive impact and the populations of some species are starting to stabilize.

That is not to say that penguin conservation is a new thing, and not all past endeavours have been successful, or particularly ethical. Probably the first and most controversial penguin conservation effort was when the Norwegian Nature Protection Society introduced several species of penguin into northern Norway, in the hope that they would colonize the fjords and islands. They started with around nine king penguins in 1936, brought back from South Georgia on a whaling ship, and later released around sixty macaroni and African penguins. The penguins soon disappeared into the ocean, and most of them seemed to have died after only a year or two. Their survival was not helped by the local people, who had not been informed of the experiment. One unfortunate king penguin wandered into a local farm and was killed by the farmer's wife who thought it was a demon. There were occasional sightings of them up and down the coast over the next ten or fifteen years, but it is unclear if breeding ever took place and the last confirmed sighting in Norway was in 1949.

Since then, things have improved, as our knowledge of how to help the birds has developed considerably. Today, penguin conservation is helping thousands of penguins around the Southern Hemisphere.

You can categorize these conservation efforts into three basic types: field conservation, science and research, and political and policy change.

It's the first one of these that tends to get most attention and is what we think of as conservation. And there have been many notable examples over the past few decades. One of the most successful examples is the amazing effort at the turn of the century to save African penguins when the ship MV *Treasure* sank off the coast of South Africa. The stricken ship was carrying 1,300 tonnes of heavy fuel oil, much of which spilt into the sea. The location of the incident could not have been worse, halfway between Dassen Island and Robben Island, which at the time constituted the largest and third-largest colonies of the critically endangered African penguin. Over the next ten days, more than 20,000 birds were washed up on the beaches covered in oil. With a large percentage of the remaining population at risk of death and a real extinction risk to the whole species, a huge conservation effort kicked in. International Bird Rescue and the International Fund for Animal Welfare (IFAW) mobilized large teams of people to collect oiled birds from the beaches, taking them back to specialized rehabilitation centres. Meanwhile, other groups of volunteers were tasked with collecting the remaining 20,000 birds from their nests, to ensure that they did not go out to sea and suffer the same fate. After the oil spill was treated and dispersed, the penguins were released and rehomed. Of the 20,000 oiled birds found on the beaches, 90% survived after treatment. It was a huge effort and an amazing success, after what could have been a true disaster for the species. Lessons learnt from shipping incidents in the previous decades had honed the response and skills of the teams working on the penguins, which contributed to the high survival rate.

It isn't always the huge international conservation efforts

that make the difference. One much more local story that I like is that of the little blue penguins at Ōamaru on South Island, New Zealand. Here, in the 1990s, penguins started to nest in an abandoned quarry near the edge of town. The local council had development plans for the land and started to fence off the site, but the town residents protested and instigated a petition to keep the penguins. The council relented and a penguin sanctuary was formed. The area was fenced off to keep stray dogs and cats away and revegetated to make the penguins feel more at home. Nest boxes were provided to encourage the birds and a monitoring programme started.

Little blue with nest box

To pay for the conservation efforts, tourists were encouraged and charged to watch the daily penguin parade. Over time, the site blossomed, penguins increased in number from 20 to around 300 pairs, and the location became one of the best in New Zealand to see little penguins. It is a really good example of how penguins and people can live in harmony and is just one of many, many efforts on a local, regional and international scale helping these endangered birds on the ground. Actions such as fencing off breeding colonies from predators, rehoming birds, and the rehabilitation of native habitat can all have a positive impact on penguin numbers.

If fencing off the predators doesn't work, more extreme measures must be taken. This is especially true for non-native invaders, against which the poor penguins often have no defence. On many sub-Antarctic islands, 'alien species' eradication projects have been undertaken by conservation agencies to remove rats, mice, cats, dogs and other predators. This has had a really positive impact on the penguins and many other seabirds that nest on those remote islands.

The second type of conservation is science and research. To protect an animal you really need to understand the threats to that species and what is driving any population change. Monitoring is often key to this. By watching and monitoring the birds, you will gain a better understanding of their behaviour and be able to detect any changes in the population. Around the world, conservation charities such as the WWF, the Global Penguin Society, Oceanites, Pew and many others on a local and international level help and facilitate researchers to study all species of penguin in easy-to-access, or remote, windswept islands, as well as the icy wastes of Antarctica.

The last, but by no means the least important, thing that

can be done to conserve penguins is to change policy and government strategy. This can include active policies like the creation of protected areas. In 2016, the world's largest Marine Protected Area was established in the Ross Sea in east Antarctica, to protect the marine life in that pristine area. Much of the evidence and reasoning for the protection came from the important colonies of Adélie and emperor penguins that live there. Other important examples include the recent fishing exclusion zones around South Africa's penguin colonies in an effort to protect the dwindling number of African penguins. In the Southern Ocean, around Antarctica, the waters are managed by the international body CCAMLR. This commission meets annually to set fishing quotas in the Southern Ocean, with the remit of ensuring that fishing does not impact the unique ecosystems of penguins, seals, seabirds and other animals, in what is one of the last near-pristine areas of the world. Its remit is to conserve marine life, using an ecological-based approach, while still enabling the harvesting of fish and krill. It also has a mandate to set up and maintain Marine Protected Areas (such as the Ross Sea Marine Protected Area). Because it is an international commission based on consensus, it doesn't always agree, and in more recent times new conservation measures have been harder to implement. But overall, it still stands out as one of the most successful international conservation bodies and is a key player in the conservation of penguins globally.

The big conservation charities that have an interest in penguins lobby for change in many political forums, where the safety of penguins and many other marine ecosystems can be an issue. Those issues can range from the management of tourist ships in Antarctica, to stopping the illegal trade in the capture and sale of penguins and other animals. Actively lobbying can help, as well as participation in events and the management bodies that help conserve the species. Setting up a fishing exclusion zone, or mitigating bycatch, could be the difference between a declining or a thriving population for some species. For other species, educating local communities might be the key to avoiding disturbance or habitat loss.

For other threats such as climate change, finding a solution is not so easy. You just cannot fix that problem at a local level and there is no regional mitigation that you can put in place to stop the ice melting or prevent the ocean currents from changing. One of the few things that you can do is tell people of the plight of these wonderful birds and the threats that they are under. Penguins, especially emperor penguins, have become a symbol of the effect of rising temperatures on the ecosystems and animals of our planet. For them, there are no other factors involved in their potential demise. It is just climate change. In a way, the best way to try to save them is to tell their own story and keep telling it, in the hope that it may make a difference and speed up our transition to a sustainable planet. It's for the benefit of not only emperors, but all penguins and wild animals, everywhere.

Penguin Jokes

Why did the penguin cross the road?
He went with the floe.

What is black and white and red all over?
A sunburnt penguin.

Why don't polar bears eat penguins?
They can't get the wrappers off.

Who is the chief of the emperor penguins?
Julius Freezer.

Who's a penguin's favourite relative?
Aunt Arctica.

How do penguins make a decision?
They flipper coin.

Why do penguins carry fish in their beaks?
Because they don't have pockets.

PENGUIN JOKES

A man walks into a pub and there's a penguin sitting at the bar. The man says to the barman, 'Why do you have a penguin at the bar?' The barman says, 'I don't know, he comes in Monday, Tuesday, Wednesday, Thursday, Friday, Saturday and Sunday.' 'Oh, that explains it,' says the man. 'He's Adélie penguin.'

There are lots more, but like the penguin on top of an iceberg said – it's all downhill from here!

Epilogue

It's November 2023 and I am out on the ice again at the Snow Hill emperor penguin colony in the Weddell Sea. I am here with a small team of scientists and field guides to count the population of the birds and to place tracking equipment on some of them to see where they go. It's my third time visiting the colony and it is good to be back. In the distance, I can hear the constant melodic fluting calls of thousands of little grey chicks, interspersed with the occasional blunter, brash trumpeting of the adults, each calling to locate its young. The sound of an emperor colony must be one of the most memorable in nature; it has a magical, haunting quality that seeps into your brain and colours your dreams.

At this time of year the chicks have reached maximum cuteness. Earlier, when they are newly born, they are scrawny and too thin to be regarded as cute, and later in the season, in December, before they fledge into their adult plumage, they become a little scruffy and lose some of that adorable charm. But now, in November, they are little chubby balls of grey fluff, with dark eyes and wonderful faces, the sort of images that you see adorning the screen in primetime documentaries and on countless Christmas cards.

EPILOGUE

We are crossing the sea ice between sub-colonies. It's dangerous going, and we check the ice regularly for cracks and thickness. The emperors breed on fast ice: the frozen sea that is attached to the land, which makes a stable platform for them to bring up their chicks. In the winter, when it is dark and extremely cold, the whole colony will huddle together in one huge group, several thousand strong, to keep warm, but now, as the weather warms, there is no need to huddle and the adults' main concern is feeding their chicks. The adults find their young by calling and, as they have left them home alone on the ice for up to a week between feeds, there is lots of time for the inquisitive youngsters to move about amongst the colony. With so many thousands of birds in a colony, and so many competing voices, it can be difficult for the adults to find their chicks, so the colony splits into sub-colonies, each with a few hundred birds, to make it easier for each parent to locate its offspring. There are about a dozen of these sub-colonies at Snow Hill Island, spread over an area of several square kilometres of ice.

Each sub-colony is a few hundred metres apart, and, as we navigate across the treacherous surface, between one group and the next, we dodge between many melt pools. The Snow Hill colony is the most northerly and probably the warmest emperor penguin colony in the world, and in the late spring sunshine temperatures can rise well above zero and melt the surface of the ice. This creates huge pools of green, scummy, algae-infested water. The year 2023 had been a warm one, and melt pools surrounded the breeding site. The chicks had hatched in early August and the continuous squirting of poo by the thousands of them over the intervening months had darkened the ice and added impurities to the snow, reducing its freezing point. It was this dirty ice, under the feet of the chicks, that melted first. Every sub-colony would have to move every few days as the snow beneath them got filthy and started to melt.

As we crossed, Steffan, our experienced ice guide, tested the surface regularly with an ice drill to ensure that it would bear our weight. We were lucky that the sea ice here was well over a metre thick, so even melt pools on the surface had not weakened its structure. Today, our second day at the colony, was cloudy and cold and the pools that had been open the day before had frozen over with a thin layer of ice. We crossed these gingerly, our boots occasionally breaking the surface and our feet breaking through into the scummy waters below. As we traversed, I kept seeing darker shapes in the water beneath this thin layer of ice on the edge of my vision. Intrigued, I tiptoed carefully over to one to investigate. It was a chick. Drowned and frozen into the green water beneath the ice. I stopped and scanned around. From where I was standing in that melt pool, I could see four or five other dark shapes, and that pool was only one of scores, perhaps hundreds, that surrounded the colony. A harrowing thought occurred to me: there would probably be several hundred chicks that had died in the pools. A grizzly end to a young life. It is easy to see how. The chicks' fluffy grey feathers are great protection against the cold, but they are useless if they get wet. The young birds are mobile and inquisitive by nature and would have had to navigate between the pools every few days when they moved locations. For a chick to fall into a melt pool would be a death sentence. It would get drenched and waterlogged and start to freeze. In November, we were still in the Antarctic spring and the sun would go below the horizon each evening, bringing with it the freezing polar night. In the darkness, as temperatures plummeted, any chick that had waterlogged feathers would turn into an ice cube.

My observations of the chicks in the pools was far from scientific and would need a lot more evidence to work out what percentage of the chicks are dying this way, but it left a nagging, uncomfortable doubt in my mind. This was made worse after we were back home and started getting data from the drone surveys and GPS tags that we had deployed. Previous studies, from colonies further south, had shown that the vast majority of adults at this time should have healthy chicks and in November, for most pairs, both adults were usually away foraging for food, leaving only a few mature birds at the colony, so there should be many more chicks than adults. Usually, in previous studies, the ratio was four or five chicks for every adult.

But this did not seem to be the case with our data. In most of the sub-colonies that we counted, we had almost as many adults as chicks. What's more, the GPS tags, which showed where the penguins were foraging and how often they came back to the colony, also showed a disturbing pattern. It was very clear from our tags which of the birds were parents – they had short regular trips that returned to the colony every few days to bring food to their young. The non-breeders, or birds that no longer had live offspring, just left the colony and didn't return. But only a quarter of the tracks were like this. Three-quarters of the adult birds we tagged were leaving the colony for good; these could only be non-breeders or adults whose chicks had died. This was not a result that we had expected at all, and it supported my fear that chick mortality, which is usually very low at emperor breeding sites, was very high in these more northerly, warmer latitudes.

Snow Hill is the most northerly of all the emperor colonies, so at the moment it is not the same as other sites. But it is an indication of what will happen to most, perhaps all, breeding sites if climate change continues unabated over the coming decades. Over the last few years, I have been documenting even worse events, when the sea ice at some emperor breeding sites had broken up way too early, causing young chicks to go into the water before they got their waterproof feathers, leading to total chick mortality, where all the young have perished. The number of sites where this is happening is increasing and in 2022, almost a third of all of the sixty-six breeding sites of emperor penguins lost the ice on which they breed before the end of the chick fledging period, with tens of thousands of chicks perishing.

Emperor penguins, and penguins in general, are at the sharp end of the effects of a warming globe. This family of birds has adapted, over the millennia, to be perfectly suited to the extreme cold, but if that environmental niche for which they are adapted no longer exists then they no longer have a place in the ecosystem, and these – some of the most charismatic beings on earth – will cease to exist. As you can see, they cannot adapt, they are slow to evolve and the human-induced changes are happening just too fast.

EPILOGUE

I am passionate about penguins and about protecting penguins. I want to get across my love for these clumsy, adorable, fluffy characters, celebrating their uniqueness and place in our world, but I am also keen to tell the sad story of our interaction with them and how we are driving many types of penguin towards extinction. The stark thing that has hit me while writing this book is the paradox between our love of these creatures and our disturbing drive towards their destruction. Communicating that message has been a hard balance to find. To tread a line between celebrating this most charismatic of birds and the often rather depressing tale of what we have done in the past, and still are doing, to them is difficult. But there is one crucial fact that we should focus on. Although several species of penguin are in real trouble, there is still hope and time to turn their fortunes around. Hopefully all eighteen (or however many it is) species will be with us for a long time to come.

I hope that you have found this a colourful, informative and interesting read (there is no lack of material on that score), while also being a sobering and thoughtful insight into the future of the penguin.

Peter Fretwell
February 2025

Penguin Books

Penguin Random House is one of the world's largest and most historic publishing companies. Its ethos is to make books accessible to everyone, promoting diversity, equality and inclusion through the power of the printed word. It publishes almost all types and genres of literature, both in paper and electronic format, across the globe. It has a huge current and back catalogue of bestsellers and famous titles, from hundreds of the world's most lauded authors, that have gained innumerable accolades over the years, including at least eighty Nobel Prizes for literature.

Penguin Books has a long and illustrious history. It started in 1934, when Allen Lane, the director of Bodley Head, at the time a small, London-based, struggling book publishing company, decided to bring out a new range of affordable but quality paperbacks. The story that unfolded has become almost legendary in the publishing world. Lane was at Exeter St David's station, looking for something to read on the long train journey back to London after visiting his friend Agatha Christie. He was appalled by the terrible choice of paperbacks available at the station bookshop, which was full of tatty, cheap, trashy novels, with gaudy covers and poor printing. He decided there and then to bring out a new and better range of paperbacks, priced as cheaply as a cigarette packet, but with the best literary titles and novels.

Stuck for a name for the range, his secretary suggested Penguin, because they were 'dignified but flippant'. He immediately sent his office assistant down to the new penguin enclosure at London Zoo to draw a penguin for the cover and soon the Penguin logo was created, an icon that has become synonymous with paperbacks for almost a century and one of the most famous company symbols in the UK. The simple, clean, double-panelled cover design assured quality, and with famous authors and titles available at an affordable price, Lane's vision was to democratize literature.

His intuition hit a nerve. Other publishing houses scoffed that Lane would never sell enough to make the range profitable, but when Woolworths and other retailers decided to sell the whole range sales boomed and within ten months of the first titles hitting the shelves in 1935 the books had sold over 1 million copies. By the mid-1930s the nation's economy was once again growing after the dark years of the Depression and the expanding literate middle classes of the interwar years were yearning and aspiring for knowledge. Affordable, quality books gave them that key and the books flew out of the shops. So successful was the range that a year later Lane, along with his brothers, started out on their own and created the Penguin Books company, specifically to sell the paperbacks. Well-known authors such as Agatha Christie, André Maurois and Ernest Hemmingway graced the covers of the original titles. Those first books were colour-coded; covers adorned in orange came to signify fiction, with blue for biography and green for crime. The original books were quality, mass-produced and sold for just a sixpence, and soon Penguin Books became a publishing sensation, monopolizing the paperback market for decades to come. In 1939, a non-fiction range called Pelican Books was launched, with pale blue covers, after Allen Lane reputedly overheard someone at a King's Cross station bookshop mistakenly ask for 'one of those Pelican books'. Well-known stars of the literary scene such as George Bernard Shaw and H.G. Wells were the first writers of the new series. Early in the war years Puffin Books was established, specifically to create something to read for the many children that were evacuated from London and the surrounding area during the Blitz. Postwar the range continued to expand, with Penguin Classics starting in 1946. A famous lawsuit in 1960 saw Penguin go on trial, charged with obscenity for publishing *Lady Chatterley's Lover*. Penguin won the lawsuit and sales of the book soared.

In more modern times Penguin has been one of the first publishing companies to embrace the digital age, with eBooks, self-publishing and digital print. In 2013, Penguin merged with the American publishing giant Random House to create the world's largest English-language publisher and it now has a truly global reach with offices in more than twenty countries. Today, the company is part of the Bertelsmann Group, with over 18,000 new titles and more than 700 million print, audio and eBooks sold annually. It has a host of famous imprints and brands on the spines of its books, over 300 in total. Each imprint is editorially and creatively independent and specializes in different literary areas.

As well as the iconic Penguin, there is Viking, Transworld, Michael Joseph, Ebury, Vintage and Puffin to name but a few. The stated mission of the group is *'to ignite a universal passion for reading by creating books for everyone. We believe that books, and the stories and ideas that they hold, have a unique capacity to connect us, change us, and carry us forwards towards a better future for generations to come.'*

In 2025 Penguin celebrates its ninetieth anniversary. Ninety years at the top of the book pile.

British Antarctic Survey

British Antarctic Survey is the UK's national polar research institution, tasked with providing science and logistics primarily for Antarctica, but also with facilitating UK research in the Arctic. Its aim is to be a *'world-leading centre for polar science and polar operations, addressing issues of global importance and helping society adapt to a changing world'*. Its headquarters are in Cambridge in the UK, but it has five research stations in Antarctica and operates the UK's primary research facility on Svalbard in the high Arctic, as well as operating a cutting-edge research ship and a fleet of five aircraft.

The survey, commonly known as BAS, has a long and distinguished history. Its origins date back to the Second World War, when, in 1943, a top-secret military operation was instigated to the Antarctic Peninsula. The operation was named after a famous Parisian nightclub called Bal Tabarin. The role of Operation Tabarin was ostensibly to deny safe anchorages to enemy raiding vessels and to gather meteorological data for allied shipping in the South Atlantic. This was accomplished by setting up a number of small bases to establish a permanent UK presence in the region. The presence also actively reinforced British territorial claims to the Peninsula that had been first established in 1908 but were at the time challenged by several South American countries.

After the war, the bases were moved away from military control but retained to ensure that a UK national presence was preserved as part of the Falkland Islands Dependencies Survey (FIDS). The main task of the now-civilian, but government-run organization, became surveying and mapping the coast and interior of what was at the time a mostly unknown continent at the bottom of the world. At its height, FIDS had 16 huts and refuges dotted around the coastline of the thousand-kilometre-long Peninsula.

In 1957/8 the Third International Polar Year saw a huge international collaboration in Antarctica and an impetus towards solving the political problems of multiple territorial claims to the continent. A year later the Antarctic Treaty was proposed and in 1961 twelve countries, including the UK, became the original signatories. The basic gist of the Treaty, leaving out the more technical language, was to put on hold any territorial claims and preserve Antarctica as a continent for peace and science. Today fifty-eight nations are parties in the Treaty system, and it is held up as an exemplar of international cooperation.

With the agreement of the Antarctic Treaty, the UK was keen to distinguish its Antarctic claim from its other UK territories in the region. As a result, in 1962, the UK renamed its Antarctic claim the British Antarctic Territory. At the same time, the British Antarctic Survey was established to continue and build on earlier scientific successes as well as act as a formal UK presence within the British Antarctic Territory.

Antarctic operation had to be separated from the Falkland Islands, with the creation of the British Antarctic

Territory, still a UK overseas territory, but now with its sovereignty rights put on hold under the Antarctic Treaty. The emphasis was now strongly on understanding the white continent, so in 1962 the British Antarctic Survey was formed with a focus on research and data gathering. With the inception of the Treaty, the British promoted their presence on the continent by focusing on science and survey. Over the years the number of research stations was rationalized, with specific stations allocated for different types of science. Rothera Station, halfway down the Antarctic Peninsula, was established in 1975 and became the hub for logistics and operations. Today it is the primary UK Antarctic facility and the only station that is occupied year-round; it remains the only station in Antarctica with both a hard runway and a deep-water wharf.

The initial scientific focus for BAS was mapping and geophysical survey as well as atmospheric, geological and biological sciences. Work was mainly gathering primary data in an area of the world that humanity knew very little about. Even when I started with the organization twenty-five years ago, there were many gaps in our knowledge of the environment in areas such as topography, ice, oceans and biology. But since then, rapid advances in technology have enabled us to expand our data gathering, and innovations such as satellites, drones, automated weather stations and autonomous oceanographic buoys have allowed us to fill many of those primary data gaps. Landmark discoveries such as the Antarctic Ozone Hole, and the extraction of deep ice cores that have proved the link between carbon dioxide and global warming, as well as science to understand and quantify the threat of global sea-level rise from Antarctica, are some of the key highlights of the many thousands of scientific papers that have been published by BAS.

With more knowledge, the science focus has shifted from primary survey to monitoring change and gathering an understanding of the Earth System. This is especially relevant in today's environment where climate change has become a driving focus for polar research. Polar regions like the Antarctic Peninsula and west Antarctica have seen rapid warming and ice loss in recent decades, so tracking these changes and understanding the drivers, physical connections and implications, and predicting future scenarios, both in Antarctica and the Arctic, is an overriding motivation for much of the research we now do.

And what happens in Antarctica does not stay in Antarctica. The majority of the world's ice sits on the frozen continent, but when that ice melts it goes into the world's oceans, and its ice caps hold enough water to raise sea levels around the globe by 58 m. Additionally, the Southern Ocean that surrounds the continent is a key driver of the world's weather systems and energy budget, so the world is very much connected to any changes to these remote and inaccessible places. With the twin global crisis of climate change and biodiversity loss, BAS's research today is centred around how the physical environment and ecology of the polar regions are responding to change and how the changes at the poles may affect humanity in the future.

But the white continent is still remote, harsh and difficult to work in. Over sixty years BAS has built up unprecedented levels of knowledge and skill in how to do work in these difficult regions. Field parties still go out into the deep field, spending time on the ice living in tents, much like the polar explorers of old once did, but now we have satellite phones and modern equipment and a much better idea of how to operate safely in such a dangerous environment. Much of the research work is collaborative, with facilities and logistics shared between international partners and the resulting data made freely available globally.

Even with all the technology and advances there are still data gaps in many areas of polar science. You may be able to see and measure the earth's surface by satellite, but you cannot see under the ice or beneath the ice shelves. Ground and aerial survey still takes place, and you still need people to go out to sample and study the animals and ecosystems on the ground, and, although BAS has many autonomous instruments, these still need to be deployed and maintained. So, BAS still has a large human presence on the continent and surrounding polar oceans, and as change becomes more rapid and the ice, the oceans and ecosystems respond, to understand and predict the consequences of those changes becomes more critical.

In my two-and-a-half decades working for BAS I have been to Antarctica six times, and I have spent time on ships, on research stations and in tents in the deep field. I have been deployed by boat, helicopter and plane and had some of the most memorable experiences of my life, but I certainly would not consider myself a field veteran. Many of the people who I regularly talk to in the organization have been 'South' over twenty times and have spent years of their lives on research stations or in tents doing fieldwork. They have collected crucial data to understand Antarctica and the Earth System and visited places where no other humans have travelled to. They have suffered hardship and isolation in some of the most difficult and most extreme environments on the planet in the pursuit of human knowledge. I am lucky enough to call these people my colleagues and friends and feel honoured and privileged to be counted amongst them.

Further Reading

Unsurprisingly, there are quite a few books about penguins. Depending upon whether you would like a heartwarming novel or a deep-dive into penguin ecology and behaviour, or specific information about a particular favourite species or location, I have tried to list some of the best and some of my personal favourites here.

For a more academic vision of the penguin family, most penguinologists would recommend Tony Williams' *The Penguins Spheniscidae*. It is often quoted as the bible for scientists. Sadly, it is currently out of print and, as it was written thirty-five years ago now, it is missing a great deal of more up-to-date information. There has not really been an equivalent scientific review since.

There have however been plenty of lighter, and slightly more accessible, reads. Many of these are coffee-table-sized hardbacks that come with a dazzling variety of photos. These include the excellent *Penguins of the World* by Wayne Lynch; *Penguins: the Ultimate Guide* by Tui De Roy, Mark Jones and Julie Cornthwaite; *Penguins: The Secret Lives of the World's Most Intriguing Birds* by Brutus Östling; and *The Nature of Penguins* by Jonathan Chester. These are all great books and have wonderful photos, often taken by the authors. Not all are still in print, but they are worth digging out if you can find them. Another popular genre is the photo guide or a book that records a personal journey to visit all eighteen penguin species. Examples of these include *Every Penguin in the World: A Quest to See Them All* by Charles Bergman and *Mission Penguin: A Photographic Quest from the Galápagos to Antarctica* by Ursula Clare Franklin.

My friend and colleague Gerald Kooyman's *Penguins: The Animal Answer Guide* is a narrative that gives answers to many questions about penguins, especially about their biology, on which Gerry is a world expert. Gerry also recently wrote a great book, almost an autobiography, about his pioneering research on emperor penguins called *Journeys with Emperors*, which is well worth a read. Another former colleague, Bernard Stonehouse, who was one of the first great penguin scientists, but is now sadly deceased, also wrote a fun and beautifully illustrated book, *A Visual Introduction to Penguins*, targeted at older children and young adults, but accessible whatever your age.

For children, there seem to be countless possibilities. I cannot say that I have read them all, but I know that my great friend Michelle LaRue's book *Young Zoologist Emperor Penguin* is a great illustrated guide to the rock star of the penguin world. But, generally, you only need to put 'penguin books for kids' into a search engine and you will be bombarded by a plethora of possible choices from the big publishing houses like National Geographic and Usborne. Some particular favourites of mine from the multitudes available are *Lost and Found* by Oliver Jeffers; *365 Penguins* by Jean-Luc Fromental and Joëlle Jolivet; and *Einstein the Penguin* by Iona Rangeley. Julia Donaldson has also recently written a penguin-themed story called *Jonty Gentoo: The Adventures of a Penguin*, beautifully illustrated by her long-time collaborator Axel Scheffler. The publication supports the work of the UK Antarctic Heritage Trust to take care of the old historic abandoned bases in Antarctica.

This leads us on to novels and stories about penguins, of which there are boundless options. Two of my recent favourites have been the darkly comic *Death and the Penguin* by Andrey Kurkov, and the lighter and warmly rewarding *The Penguin Lessons* by Tom Michell, a true story now transformed into a major film. Both of these novels are about individuals who adopt a penguin, but in very different circumstances. Hazel Prior has written a series of uplifting penguin-themed books, *Gone with the Penguins*, *Call of the Penguins*, *Away with the Penguins* and *How Penguins Saved Veronica*, which shows just how loved these birds are. Then of course there is the original penguin novel *Mr Popper's Penguins*, penned by Richard Atwater and Florence Atwater way back in 1938, but just as good today.

There are a couple of specific books about the uplifting and miraculous true story of the rescue of thousands of penguins from an oil spill off the coast of Africa. Both are called *The Great Penguin Rescue*; take your pick from Dyan deNapoli's or the version by Sandra Markle.

And there are many others, including more general bird books such as Stephen Moss's *Ten Birds that Changed the World*. Apologies if I have missed a personal favourite or real nugget of penguin wisdom, but as you can see there is no shortage of material for you to dive into if you would like to become a real penguin aficionado. Happy reading!

Glossary

Aptenodytes – A genus containing the emperor and king penguins.
Archipelago – A group of islands.
Auricular patches – Soft feathers on a bird's ears that protect them and reduce wind noise.
Austral summer – The period from December to February in the southern hemisphere.
Ballast – A weight.
Banded penguin – A penguin of the *Spheniscus* genus with a black stripe and/or other markings around the top half of its body.
Bioluminescence – An organism that can emit light from its body in water or in the air.
Booby – A seabird. The name originates from the Spanish word *bobo*, meaning foolish.
Brood – A group of chicks that are born at the same time.
Brush-tailed – A genus of penguins scientifically termed *Pygoscelis*, namely, Adélie, chinstrap and gentoo.
Bycatch – Unintentional capture of birds, fish, whales, dolphins or turtles while fishing in the sea.
Chest bib – A white patch on the chest of a black bird.
Colony – A large group of birds living together.
Crested penguins – Penguins of the *Eudyptes* genus, black-and-white-penguins with yellow crests and red bills and eyes, found on sub-Antarctic Islands.

GLOSSARY

Cromwell Current – A current of oceanic water in the Pacific that flows east along the equator.

Crustaceans – Crabs, lobsters and crayfish.

Dive logger – A device that collects information about the diving behaviour of penguins or seals.

Downy feathers – Very soft, insulating feathers that can be found on chicks and adult penguins, with fine fluffy filaments that are excellent at insulating against the cold.

El Niño – A climatic change, mainly in the equatorial regions of the Pacific Ocean, that brings unseasonally warm ocean temperatures and currents to western South America and often has other meteorological effects throughout the Pacific region.

Embayment – A recess in a coastline forming a bay.

Endemic – A bird that is restricted to a particular country or region.

Eudyptes – A genus of penguin commonly known as the crested penguins.

Eudyptula – A genus of penguin that currently contains one species, the little penguin.

Eurocentric – A focus on Europe or Europeans.

Fast ice – Sea ice that is attached to the coastline, sea floor or static icebergs.

Foraging – How an animal acquires food.

Fumarole – A vent in the earth's crust that releases volcanic gases and vapours.

Galapagos hawk – A predatory bird that lives on the Galapagos Islands.

Gene sequencing – A scientific method used to determine the genetic make-up of a specific organism.

GLOSSARY

GPS (Global Positioning System) transmitter – A tag that is attached temporarily to an animal to track its position and movements.

Great penguins – The two great penguin species, namely kings and emperors.

Humboldt Current – A cold, low-salt ocean current that flows along the western coast of South America in the Pacific Ocean.

Ice cap – An extensive thick layer of ice covering an area of land.

Indigenous – Native to a particular area or country.

Katabatic wind – Cold, dry, powerful winds that flow downhill from cold areas.

La Niña – A period of cooler-than-normal sea surface temperatures, primarily in the Pacific Ocean, that are associated with oceanic currents and weather systems. It is the opposite of El Niño.

Mandibular plate – A bony plate on the lower beak of a bird. For penguins this plate is often orange or pink-coloured.

Mark-recapture – A method used to estimate the size of a population of birds or animals.

Mechanical weighbridge – A non-invasive weighing scale that measures and records the weight of animals as they move across it.

Megadyptes – The genus of the yellow-eyed penguin.

Mohawk – A 1970s hairstyle featuring long hair in the middle of the head, with shaved sides.

Monitor lizard – A lizard native to Africa, Asia and Oceania.

Moulting – The process of birds replacing their feathers. For penguins this is an annual event that takes several weeks and has to be done on land or on the ice.

GLOSSARY

Nape – The back of the neck.

Occipital crest – A plume on top of a bird's head.

Overwintering – Staying in a place over winter.

Panchromatic – A term used to identify greyscale satellite images; these are often collected at a higher resolution than colour images.

Penguin digester – A machine used to remove the oil from penguins on Macquarie Island, between 1890 and 1919.

Plume – A feather or group of feathers, often long and sometimes used for display purposes.

Polar Front – An oceanic area where cool, northward-moving polar waters meet warmer, temperate waters moving south. At the front, the water temperature can change rapidly over a few kilometres. The mixing and upwelling associated with the Polar Front results in extremely productive waters.

Porpoising – A behaviour where fast-moving penguins, swimming near the surface, rhythmically leap out of the water. It is often seen close to shore and is thought to be primarily a defence mechanism against predators.

Pygoscelis – A genus of penguins commonly called the brush-tailed penguins, namely: Adélie, chinstrap and gentoo.

Quill – The hollow shaft of a feather.

Radiocarbon dating – A technique that can work out how old organic matter is. It can be used to identify the age of dead animals.

Scrape – A shallow depression in the ground where birds lay their eggs.

Striated caracara – A bird of prey that lives on the Falkland Islands and Tierra del Fuego.

GLOSSARY

Sub-Antarctic islands – Small islands in the Southern Ocean surrounding the Antarctic continent.

Sympatric – Breeding close to or amongst others.

Tasmanian devil – The largest surviving carnivorous marsupial, living on the island of Tasmania and in Australia.

Telemetry – A technique that uses wireless technology to remotely monitor animals and their environments, usually from a device attached to the animal that transmits data to a satellite or other receiver.

Time-depth recorder – A device that is attached to a marine species to record its movements in water.

Victuals – Any substance that can be used as food.

Viscosity – A fluid's resistance to flow, or the thickness of a fluid.

White-bellied sea eagle – A large bird of prey that lives in coastal and near-coastal areas of Asia and Australasia.

Acknowledgements

Lisa and I would like to thank Tom Killingbeck, who has always been enthusiastic, dedicated and supportive throughout. This book came from your mind, Tom; I hope that the result does justice to that idea.

From my work on penguins, I really need to acknowledge and thank many of the researchers who have helped me transform from a geographer who knew very little of penguins when I joined British Antarctic Survey, to what I am today. Originally, I was mentored by Phil Trathan at BAS, who encouraged and supported me in my research. I thank Richard Phillips and Norman Ratcliffe, two of the UK's leading seabird experts, for informative discussions over the years. I also have to mention Adrian Fox and Andrew Fleming, who saw the impact the work on penguins was having and enabled me to prioritize it to help it reach its full potential. There are also a number of experts at BAS who have kindly looked over the manuscript and given some helpful suggestions and comments, including Kevin Hughes and Director of Innovation and Impact Beatrix Schlarb-Ridley.

I have also worked with some great international experts: Gerry Kooyman has always been inspirational and supportive, and I have enjoyed my many penguin discussions with Barbara Wienecke. I cannot end this list without a big mention to Michelle LaRue and Stephanie Jenouvrier, whom I have collaborated with intensively over the last fifteen years. I would also like to acknowledge WWF, especially Rod Downie, who has helped fund our science for over a decade.

Lastly, I would like to thank all of the people I work with at BAS, from my own Wildlife from Space team and the staff of the Mapping and Geographic Information Centre, to the media department and everyone else who has supported me in my journey to open a window on the lives, locations and future of penguins.

Adélies jumping into the water

Chinstrap scratching

Gentoo chicks asking to be fed

Adélie side stare

Humboldt chick tentatively entering water

African (Jackass) penguin braying

Fiordland penguin rock hopping

Emperor penguin chick cooling down on the snow